W9-CCI-169

QuickStudy®

for

Chemistry

Boca Raton, Florida

DISCLAIMER:

This QuickStudy® Booklet is an outline only, and as such, cannot include every aspect of this subject. Use it as a supplement for course work and textbooks. BarCharts, Inc., its writers, editors and designers are not responsible or liable for the use or misuse of the information contained in this booklet.

©2006 BarCharts, Inc.
ISBN 13: 9781423202615
ISBN 10: 1423202619

BarCharts® and QuickStudy® are registered trademarks of BarCharts, Inc.

Author: Mark D. Jackson, Ph.D.
Publisher:

      BarCharts, Inc.

      6000 Park of Commerce Boulevard, Suite D

      Boca Raton, FL 33487

      www.quickstudy.com

Printed in Thailand

Contents

Study Hints

NOTE TO STUDENT:

Use this QuickStudy® booklet to make the most of your studying time.

QuickStudy® samples offer easy-to-understand explanations and step-by-step instructions.

SAMPLE:
a 0.01 M solution of H_3O^+ has a pH of 2

QuickStudy® tips alert you to common studying pitfalls; refer to them often to avoid problems.

⚠ **Pitfall:** Your calculator has separate keys for lnx (base e), logx (base 10), 10^x and e^{xf}.

QuickStudy® notes provide need-to-know information; read them carefully to better understand key concepts.

NOTE:
The change "a" is the same for each because of the 1:1:1:1 coefficients in the equation.

Take your learning to the next level with QuickStudy®!

The ABCs of Chemistry

Nomenclature

■ **Chemical Names**
Inorganic: Start with the "cation" name followed by "anion"; use prefixes to clarify any ambiguity. Anion, cation names derived from element names.
Organic: Separate naming system.

■ **Chemical Formulae**
◆ Cation symbol, followed by anion
◆ Subscripts denote relative composition
◆ Enclose polyatomic ions or molecules in parentheses
◆ **Molecular Formula:** Discrete molecule
◆ **Empirical Formula:** Relative molar ratio of elements for solids or molecules

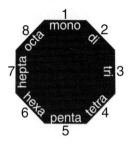

■ Monatomic Cations

Alkali metal (+1); alkaline earth (+2);
Transition metal: valence varies, give the valence in the name and formula; "ous" or "ic" ending.
Roman-numerals are less confusing, more general.

ferric, Fe(III)	ferrous, Fe(II)
stannic, Sn(IV)	stannous, Sn (II)
plumbic, Pb(IV)	plumbous, Pb(II)
cupric, Cu(II)	cuprous, Cu(I)
mercuric, Hg(II)	mercurous, Hg(I)

■ (+1) Polyatomic Cations

ammonium, NH_4^+
hydronium, H_3O^+; active form of acid in water

■ Monatomic Anions

(-4) carbide, C^{4-}, silicide, Si^{4-}
(-3) nitride, N^{3-}, phosphide, P^{3-}, arsenide, As^{3-}
(-2) oxide, O^{2-}, sulfide, S^{2-}, selenide, Se^{2-}, telluride, Te^{2-}
(-1) hydride, H^- halides: fluoride F^-, chloride Cl^-, bromide, Br^-, iodide, I^-
     Acids: hydro -fluoric, -chloric, -bromic, -iodic

■ Polyatomic Anions (and Respective Acids)

(-1)

acetate, $C_2H_3O_2^-$	*acetic acid, $C_2H_4O_2$*
nitrate, NO_3^-	*nitric acid, HNO_3*
nitrite, NO_2^-	*nitrous acid, HNO_2*
hypochlorite, ClO^-	*hypochlorous acid, HClO*
chlorite, ClO_2^-	*chlorous acid, $HClO_2$*
chlorate, ClO_3^-	*chloric acid, $HClO_3$*
perchlorate, ClO_4^-	*perchloric acid, $HClO_4$*
cyanide, CN^-	*hydrocyanic acid, HCN*
hydroxide, OH^-	*formed by bases in water*

bicarbonate or hydrogen carbonate, HCO_3^-
bisulfate or hydrogen sulfate, HSO_4^-
dihydrogen phosphate, $H_2PO_4^-$
permanganate, MnO_4^-

■ (-2)
carbonate, CO_3^{2-}	carbonic acid, H_2CO_3
sulfate, SO_4^{2-}	sulfuric acid, H_2SO_4
sulfite, SO_3^{2-}	sulfurous acid, H_2SO_3
chromate, CrO_4^{2-}	chromic acid, H_2CrO_4

peroxide O_2^{2-}
biphosphate or hydrogen phosphate, HPO_4^{2-},
dichromate, $Cr_2O_7^{2-}$
thiosulfate, $S_2O_3^{2-}$ (thio: S substitute for O atom)
disulfide, S_2^{2-}

■ (-3)
phosphate, PO_4^{3-} phosphoric acid, H_3PO_4

■ (-4)
silicate, SiO_4^{4-} silicic acid, H_4SiO_4

Measurement & Units

mass: Kilogram (kg) = 1000 g = 2.2046 pound
length: Meter (m) = 100 cm = 1.0936 yard = 1010 Å
time: Second (s);
temperature: Kelvin, (K)

$$T(K) = T(°C) + 273.15$$
$$T(°F) = \frac{9}{5} \times T(°C) + 32$$

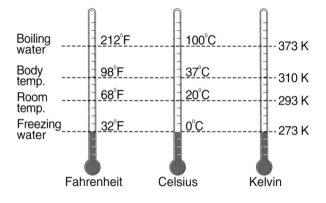

	Fahrenheit	Celsius	Kelvin
Boiling water	212°F	100°C	373 K
Body temp.	98°F	37°C	310 K
Room temp.	68°F	20°C	293 K
Freezing water	32°F	0°C	273 K

volume: liter (L) = 1000mL = 1.0567 quart
pressure: Pascal, Pa (N/m^2); 1 atm = 101,325 Pa
force: Newton, N (J/m)
charge: coulomb, C
energy: Joule, J ($kg\ m^2/s^2$) = 0.23901 calorie

Prefixes:

tera T (10^{12})	**giga** G (10^9)	**mega** M (10^6)
kilo k (10^3)	**centi** c (10^{-2})	**milli** m (10^{-3})
micro μ (10^{-6})	**nano** n (10^{-9})	**pico** p (10^{-12})

Fundamental Constants

R = 8.314 J $mole^{-1}$ K^{-1} (for energy calculation)
R = 0.082 1 atm $mole^{-1}$ K^{-1} (for gas calculation)
Avogadro's Number: N_A = 6.022 x 10^{23} $mole^{-1}$
Boltzmann constant:
      k = R/N_A = 1.381 x 10^{-23} J $molecule^{-1}$ K^{-1}
Elementary charge of the electron, e:
      1.602 x 10^{-19} C
Faraday Constant, $\mathcal{F}$: charge of N_a electrons.
Mass of a proton, m_p: 1.673 x 10^{-27} kg
Mass of a neutron, m_n: 1.675 x 10^{-27} kg
Mass of an electron, m_e: 9.110 x 10^{-31} kg
Planck's Constant, h: 6.626 x 10^{-34} J s
Speed of light in a vacuum, c: 2.9979 x 10^8 m s^{-1}

2 The Elements

Periodic Table of the Elements

Key:
- 1 — ATOMIC NUMBER
- H — SYMBOL
- 1.008 — ATOMIC WEIGHT

1	2	3	4	5	6	7	8	9	10	11	12	13	14	15	16	17	18
H 1 1.01																	He 2 4.00
Li 3 6.94	Be 4 9.01											B 5 10.81	C 6 12.01	N 7 14.01	O 8 16.00	F 9 19.00	Ne 10 20.18
Na 11 22.99	Mg 12 24.31											Al 13 26.98	Si 14 28.09	P 15 30.97	S 16 32.07	Cl 17 35.45	Ar 18 39.95
K 19 39.10	Ca 20 40.08	Sc 21 44.96	Ti 22 47.88	V 23 50.94	Cr 24 52.00	Mn 25 54.94	Fe 26 55.85	Co 27 58.93	Ni 28 58.69	Cu 29 63.55	Zn 30 65.39	Ga 31 69.72	Ge 32 72.61	As 33 74.92	Se 34 78.96	Br 35 79.90	Kr 36 83.80
Rb 37 85.47	Sr 38 87.62	Y 39 88.91	Zr 40 91.22	Nb 41 92.91	Mo 42 95.94	Tc 43 97.91	Ru 44 101.07	Rh 45 102.91	Pd 46 106.42	Ag 47 107.87	Cd 48 112.41	In 49 114.82	Sn 50 118.71	Sb 51 121.75	Te 52 127.60	I 53 126.90	Xe 54 131.29
Cs 55 132.91	Ba 56 137.33	La* 57 138.91	Hf 72 178.49	Ta 73 180.95	W 74 183.85	Re 75 186.21	Os 76 190.20	Ir 77 192.22	Pt 78 195.08	Au 79 196.97	Hg 80 200.59	Tl 81 204.38	Pb 82 207.20	Bi 83 208.98	Po 84 208.98	At 85 209.99	Rn 86 222.02
Fr 87 223.02	Ra 88 226.03	Ac* 89 227.03	Rf 104 261	Db 105 262	Sg 106 263	Bh 107 262	Hs 108 265	Mt 109 266	Ds 110 281	111 272	112 277	Uut 113 284	Uuq 114 289	Uup 115 288	116 289	117	118 293

Discovery notes: Ds 110 (Dec. 1994), 111 (Dec. 1994), 112 (Dec. 1996), Uuq 114 (Dec. 1999); 116, 117, 118 — Not Discovered.

La* 57	Ce 58 140.12	Pr 59 140.91	Nd 60 144.24	Pm 61 144.91	Sm 62 150.36	Eu 63 151.97	Gd 64 157.25	Tb 65 158.93	Dy 66 162.50	Ho 67 164.93	Er 68 167.26	Tm 69 168.93	Yb 70 173.04	Lu 71 174.97
Ac* 89	Th 90 232.04	Pa 91 231.04	U 92 238.05	Np 93 237.05	Pu 94 244.06	Am 95 243.06	Cm 96 247.07	Bk 97 247.07	Cf 98 251	Es 99 252.08	Fm 100 257.10	Md 101 258.10	No 102 259.10	Lr 103 260.11

5

Characteristics of Elements 1-54

ATOMIC NUMBER	SYMBOL	ATOMIC WEIGHT	NAME	ELECTRON CONFIG.
1	H	1.008	Hydrogen	$1s^1$ **n=1**
2	He	4.003	Helium	$1s^2$
3	Li	6.941	Lithium	$1s^2 2s^1$
4	Be	9.012	Beryllium	$1s^2 2s^2$
5	B	10.81	Boron	$[He]2s^2 2p^1$
6	C	12.01	Carbon	$[He]2s^2 2p^2$ **n=2**
7	N	14.01	Nitrogen	$[He]2s^2 2p^3$
8	O	16.00	Oxygen	$[He]2s^2 2p^4$
9	F	19.00	Fluorine	$[He]2s^2 2p^5$
10	Ne	20.18	Neon	$[He]2s^2 2p^6$
11	Na	22.99	Sodium	$[Ne]3s^1$
12	Mg	24.31	Magnesium	$[Ne]3s^2$
13	Al	26.98	Aluminum	$[Ne]3s^2 3p^1$
14	Si	28.09	Silicon	$[Ne]3s^2 3p^2$ **n=3**
15	P	30.97	Phosphorus	$[Ne]3s^2 3p^3$
16	S	32.07	Sulfur	$[Ne]3s^2 3p^4$
17	Cl	35.45	Chlorine	$[Ne]3s^2 3p^5$
18	Ar	39.95	Argon	$[Ne]3s^2 3p^6$
19	K	39.10	Potassium	$[Ar]4s^1$
20	Ca	40.08	Calcium	$[Ar]4s^2$
21	Sc	44.96	Scandium	$[Ar]3d^1 4s^2$
22	Ti	47.88	Titanium	$[Ar]3d^2 4s^2$
23	V	50.94	Vanadium	$[Ar]3d^3 4s^2$
24	Cr	52.00	Chromium	$[Ar]3d^5 4s^1$
25	Mn	54.94	Manganese	$[Ar]3d^5 4s^2$
26	Fe	55.85	Iron	$[Ar]3d^6 4s^2$
27	Co	58.93	Cobalt	$[Ar]3d^7 4s^2$ **n=4**
28	Ni	58.69	Nickel	$[Ar]3d^8 4s^2$
29	Cu	63.55	Copper	$[Ar]3d^{10} 4s^1$
30	Zn	65.39	Zinc	$[Ar]3d^{10} 4s^2$
31	Ga	69.72	Gallium	$[Ar]3d^{10} 4s^2 4p^1$
32	Ge	72.61	Germanium	$[Ar]3d^{10} 4s^2 4p^2$
33	As	74.92	Arsenic	$[Ar]3d^{10} 4s^2 4p^3$
34	Se	78.96	Selenium	$[Ar]3d^{10} 4s^2 4p^4$
35	Br	79.90	Bromine	$[Ar]3d^{10} 4s^2 4p^5$
36	Kr	83.80	Krypton	$[Ar]3d^{10} 4s^2 4p^6$
37	Rb	85.47	Rubidium	$[Kr]5s^1$
38	Sr	87.62	Strontium	$[Kr]5s^2$
39	Y	88.91	Yttrium	$[Kr]4d^1 5s^2$
40	Zr	91.22	Zirconium	$[Kr]4d^2 5s^2$
41	Nb	92.91	Niobium	$[Kr]4d^4 5s^1$
42	Mo	95.94	Molybdenum	$[Kr]4d^5 5s^1$
43	Tc	98	Technetium	$[Kr]4d^5 5s^2$
44	Ru	101.1	Ruthenium	$[Kr]4d^7 5s^1$
45	Rh	102.9	Rhodium	$[Kr]4d^8 5s^1$ **n=5**
46	Pd	106.4	Palladium	$[Kr]4d^{10}$
47	Ag	107.9	Silver	$[Kr]4d^{10} 5s^1$
48	Cd	112.4	Cadmium	$[Kr]4d^{10} 5s^2$
49	In	114.8	Indium	$[Kr]4d^{10} 5s^2 5p^1$
50	Sn	118.7	Tin	$[Kr]4d^{10} 5s^2 5p^2$
51	Sb	121.8	Antimony	$[Kr]4d^{10} 5s^2 5p^3$
52	Te	127.6	Tellurium	$[Kr]4d^{10} 5s^2 5p^4$
53	I	126.9	Iodine	$[Kr]4d^{10} 5s^2 5p^5$
54	Xe	131.3	Xenon	$[Kr]4d^{10} 5s^2 5p^6$

Characteristics of Elements 55-106

ATOMIC NUMBER	SYMBOL	ATOMIC WEIGHT	NAME	ELECTRON CONFIG.
55	Cs	132.9	Cesium	$[Xe]6s^1$
56	Ba	137.3	Barium	$[Xe]6s^2$
57	La	138.9	Lanthanum	$[Xe]5d^16s^2$
58	Ce	140.1	Cerium	$[Xe]4f^15d^16s^2$
59	Pr	140.9	Praseodymium	$[Xe]4f^36s^2$
60	Nd	144.2	Neodymium	$[Xe]4f^46s^2$
61	Pm	145	Promethium	$[Xe]4f^56s^2$
62	Sm	150.4	Samarium	$[Xe]4f^66s^2$
63	Eu	152.0	Europium	$[Xe]4f^76s^2$
64	Gd	157.3	Gadolinium	$[Xe]4f^75d^16s^2$
65	Tb	158.9	Terbium	$[Xe]4f^96s^2$
66	Dy	162.5	Dysprosium	$[Xe]4f^{10}6s^2$
67	Ho	164.9	Holmium	$[Xe]4f^{11}6s^2$
68	Er	167.3	Erbium	$[Xe]4f^{12}6s^2$
69	Tm	168.9	Thulium	$[Xe]4f^{13}6s^2$
70	Yb	173.0	Ytterbium	$[Xe]4f^{14}6s^2$
71	Lu	175.0	Lutetium	$[Xe]4f^{14}5d^16s^2$
72	Hf	178.5	Hafnium	$[Xe]4f^{14}5d^26s^2$
73	Ta	180.9	Tantalum	$[Xe]4f^{14}5d^36s^2$
74	W	183.8	Tungsten	$[Xe]4f^{14}5d^46s^2$
75	Re	186.2	Rhenium	$[Xe]4f^{14}5d^56s^2$
76	Os	190.2	Osmium	$[Xe]4f^{14}5d^66s^2$
77	Ir	192.2	Iridium	$[Xe]4f^{14}5d^76s^2$
78	Pt	195.1	Platinum	$[Xe]4f^{14}5d^96s^1$
79	Au	197.0	Gold	$[Xe]4f^{14}5d^{10}6s^1$
80	Hg	200.6	Mercury	$[Xe]4f^{14}5d^{10}6s^2$
81	Tl	204.4	Thallium	$[Xe]4f^{14}5d^{10}6s^26p^1$
82	Pb	207.2	Lead	$[Xe]4f^{14}5d^{10}6s^26p^2$
83	Bi	209	Bismuth	$[Xe]4f^{14}5d^{10}6s^26p^3$
84	Po	209	Polonium	$[Xe]4f^{14}5d^{10}6s^26p^4$
85	At	210	Astatine	$[Xe]4f^{14}5d^{10}6s^26p^5$
86	Rn	222	Radon	$[Xe]4f^{14}5d^{10}6s^26p^6$
87	Fr	223	Francium	$[Rn]7s^1$
88	Ra	226.0	Radium	$[Rn]7s^2$
89	Ac	227.0	Actinium	$[Rn]6d^17s^2$
90	Th	232.0	Thorium	$[Rn]6d^27s^2$
91	Pa	231	Protactinium	$[Rn]5f^26d^17s^2$
92	U	238.0	Uranium	$[Rn]5f^36d^17s^2$
93	Np	237.0	Neptunium	$[Rn]5f^46d^17s^2$
94	Pu	244	Plutonium	$[Rn]5f^67s^2$
95	Am	243	Americium	$[Rn]5f^77s^2$
96	Cm	247	Curium	$[Rn]5f^76d^17s^2$
97	Bk	247	Berkelium	$[Rn]5f^97s^2$
98	Cf	251	Californium	$[Rn]5f^{10}7s^2$
99	Es	252	Einsteinium	$[Rn]5f^{11}7s^2$
100	Fm	257	Fermium	$[Rn]5f^{12}7s^2$
101	Md	258	Mendelevium	$[Rn]5f^{13}7s^2$
102	No	259	Nobelium	$[Rn]5f^{14}7s^2$
103	Lr	260	Lawrencium	$[Rn]5f^{14}6d^17s^2$
104	Rf	261	Rutherfordium	$[Rn]5f^{14}6d^27s^2$
105	Db	262	Dubnium	$[Rn]5f^{14}6d^37s^2$
106	Sg	263	Seaborgium	$[Rn]5f^{14}6d^47s^2$

n=5

n=6

Subshells & Orbitals

s - subshell
1 orbital

p - subshell
3 orbitals

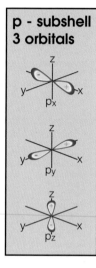

d - subshell
5 orbitals

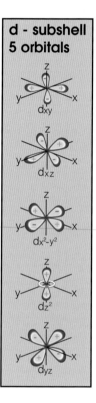

3 Atoms

Atomic Structure

■ **Atomic Number, Z:** Number of protons in the nucleus; for neutral atom, Z = Number of electrons; for anion of charge -q, Z+q electrons; for cation of charge +q, Z-q electrons.

■ **Mass Number, A:** A= Z + N, the number of neutrons in the nucleus.

■ **Isotopes:** atoms with the same Z, different A. An element can have a number of isotopes; any sample contains a number of isotopes; in practical work use the **average** Mass Number or **Atomic Weight** (weighted average of natural isotopes for a given element).

Atomic Quantum Numbers & Orbitals

■ *n*: principle
■ *l*: angular momentum (orbital shape)
■ *m_l*: magnetic (orbital direction)
■ *m_s*: electron spin
For each *n*, the possible *l* values are $0,1...n\text{-}1$; for each *l* the possible m_l values are $-l, .0..+l$; m_s has two possible values: $+\frac{1}{2}$ and $-\frac{1}{2}$ (spin up, spin down). Each value of *n* denotes a "shell" in the atomic structure; each *l* denotes a subshell.

- **s-orbital:** $l = 0$, 1 type of orbital
- **p-orbital:** $l = 1$, 3 types of orbitals
- **d-orbital:** $l = 2$, 5 types of orbitals
- **f-orbital:** $l = 3$, 7 types of orbitals

Many-Electron Atoms

Orbitals and quantum number's derived for the hydrogen atom are used to describe all many-electron atoms and ions. A more rigorous treatment determines exact energy levels and orbital properties. The main result is that sub-shells have different energies. The filling of levels is guided by the **Aufbau Principle:**

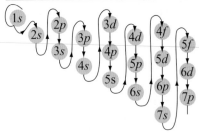

Aufbau order for filling sublevels

n	l	m_l	m_s	element	
1	0	0	$+\frac{1}{2}$	**H**	1s-orbital
1	0	0	$-\frac{1}{2}$	**He**	spherical
2	0	0	$+\frac{1}{2}$	**Li**	2s-orbital
2	0	0	$-\frac{1}{2}$	**Be**	spherical
2	1	-1	$+\frac{1}{2}$	**B**	2p-orbitals
2	1	0	$+\frac{1}{2}$	**C**	
2	1	+1	$+\frac{1}{2}$	**N**	bimodal
2	1	-1	$-\frac{1}{2}$	**O**	p_x p_y p_z
2	1	0	$-\frac{1}{2}$	**F**	
2	1	+1	$-\frac{1}{2}$	**Ne**	

■ **Electron Configuration:** Atomic orbital occupancy
■ **Pauli Exclusion Principle:** Each electron has a unique set of quantum numbers. An orbital may hold up to two electrons, one with spin-up, $m_s = +\frac{1}{2}$ and one with spin-down, $m_s = -\frac{1}{2}$
■ **Hund's Rule:** Electrons fill p, d and f sub-shells in a manner to maximize the number of unpaired electrons (half-filled orbitals)
■ **Ionization Potential (IP):** Energy required to remove electron from an atom or ion
■ **First IP:** Removal of first valence electron
■ **Electronegativity:** Tendency of an atom to attract electrons in a chemical bond; 0-4 Pauling scale, 0 for rare gases

4 Molecules

Molecular Properties

▮ **Geometry: Valence Shell Electron Pair Repulsion Theory (VSEPR):** Optimum arrangement of bonded neighbors (X) and non-bonded or lone pairs (E). All structures are derived from ideal arrangement of X+E objects about central atom.

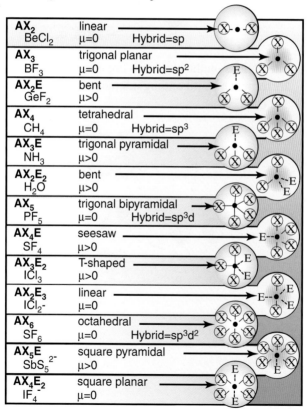

AX$_2$ BeCl$_2$	linear $\mu=0$ Hybrid=sp	
AX$_3$ BF$_3$	trigonal planar $\mu=0$ Hybrid=sp^2	
AX$_2$E GeF$_2$	bent $\mu>0$	
AX$_4$ CH$_4$	tetrahedral $\mu=0$ Hybrid=sp^3	
AX$_3$E NH$_3$	trigonal pyramidal $\mu>0$	
AX$_2$E$_2$ H$_2$O	bent $\mu>0$	
AX$_5$ PF$_5$	trigonal bipyramidal $\mu=0$ Hybrid=sp^{3}d	
AX$_4$E SF$_4$	seesaw $\mu>0$	
AX$_3$E$_2$ ICl$_3$	T-shaped $\mu>0$	
AX$_2$E$_3$ ICl$_2^-$	linear $\mu=0$	
AX$_6$ SF$_6$	octahedral $\mu=0$ Hybrid=sp^{3}d^2	
AX$_5$E SbS$_5^{2-}$	square pyramidal $\mu>0$	
AX$_4$E$_2$ IF$_4^-$	square planar $\mu=0$	

Molecular Orbital (MO) Theory

■ MO's give the Ψ for a molecule; each MO is a linear combination (weighted sum) of AO's. Overlap of atomic orbital lobes with the same sign gives constructive interference **(bonding interaction)**; overlap of orbital lobes with opposite signs gives destructive interference: a less stable arrangement termed **antibonding.**

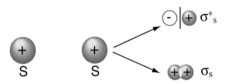

Interactions on the bond axis are termed σ; off axis are termed π. Antibonding interactions are noted π^* and σ^*.

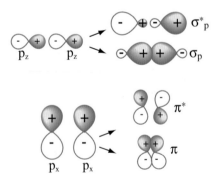

The MO's describe the electron density and energy. The bonding interaction produces an accumulation of electron density in the bonding region.

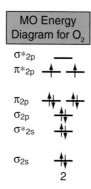

A diagram of MO's gives energy levels for electrons in the molecule. As for AO's, each MO can have at most <u>2 electrons (with paired spin)</u>.

The **bond order** is given by (number of bonding electron pairs) - (number of antibonding electron pairs).

Hybrid Orbitals

■ S,p,d and f AO's can mix or **hybridize** to form equivalent lone pairs and bonding pairs orbitals. Supports VSEPR model. S and p can hybridize to sp (2), sp^2 (3) or sp^3 (4 equivalent orbitals); d orbitals expand the options to five (sp^{3}d) and six bonded neighbors (sp^{3}d^2).

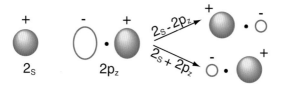

Matter & Processes

Types of Matter

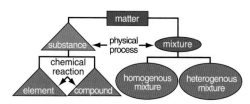

Reactions

■ **formation:** Elements form a compound:
$C + 2 H_2 => CH_4$

■ **combination (synthesis):** Two substances form a new substance: $2 Na + Cl_2 => 2 NaCl$

■ **oxidation-reduction:** Oxidation states change [*see* chapter 10 for more details]

■ **acid-base neutralization:** Form water and a salt [*see* chapter 12 for more details]

■ **decomposition:** One substance yields two or more substances: $2 HgO => 2 Hg + O_2$

■ **displacement:** Element displaces another element from a compound: $Zn + 2 HCl => H_2 + ZnCl_2$

■ **double displacement or metathesis:** Exchange anions, form precipitate:
$NaCl(aq) + AgNO_3(aq) => AgCl(ppt) + NaNO_3(aq)$

■ **combustion:** Exothermic reaction with oxygen:
$C + O_2 => CO_2$

Physical Processes

- **melting** (s => l); **freezing** (l => s)
  (at 1 atm: normal melting point "T_m")
- **evaporation** (l => g); **condensation** (g => l)
  (at 1 atm, normal boiling point "T_b")
- **sublimation** (s => g)
- **solution:** mixtures of gases, liquids or solids
- **distillation:** Separate a mixture of liquids by selective evaporation
- **triple point, T_t:** solid, liquid and gas in equilibrium
- **vapor pressure:** The partial pressure of the gas in equilibrium with liquid (depends on temperature); equals 1 atm at T_b

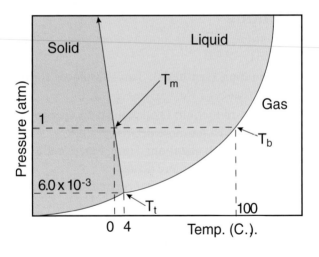

6 Chemical Interactions

■ **Electronic Properties:**

◆ **dipole moment (μ):** Position of the molecular electron density, relative to the center of mass; imparts partial charge to the molecule: $\mu = 0$ for **symmetric molecule (H_2).**

◆ **$\mu > 0$ for asymmetric molecule:**

$$^{\delta^+}H - Cl^{\delta^-}$$

◆ **polarizability:** Tendency of electron cloud to distort from equilibrium due to external electro-statics; increases with atom size

■ **Intermolecular (between molecules)**

◆ **London forces (dispersion):** Attraction of induced-dipole moments; stronger for more polarizable species; accounts for liquefaction of gases like argon and methane.

◆ **dipole-dipole:** Molecules with dipole moments experience attractive forces for certain relative orientations. Based on electrostatic forces.

◆ **hydrogen-bonding:** Enhanced dipole-interaction between hydrogen of an -OH or -NH group and a nearby oxygen or nitrogen atom

◆ **electrostatic:** Strong interaction between ions; attractive (opposite charges) or repulsive (like charges). Inversely proportional to distance and dielectric constant of the medium. Water has a large dielectric constant, stabilizes ion formation in solution.

■ **Chemical Bonds (between atoms in molecules)**
  ◆ **valence electrons:** The outer electrons which form chemical bonds. The rest of the electrons (inner-shell) are inert core electrons. Bonding is described with **three ideal** models:
  ◆ **covalent bond:** Electrons are shared; the polarity of the bond denotes unequal sharing and is determined by the electronegativity of the atoms.
    • pure covalent: equal sharing of electrons
    • polar covalent: unequal sharing of electrons
  ◆ **ionic bond:** Electrostatic interactions between ions; created by the transfer of electrons between atoms to create ions with filled valence shells.
  ◆ **metallic bond:** Electrons are delocalized: shared by a large number of metallic nuclei.
  ◆ **real bonds:** In common molecules bonds tend to be polar covalent partially covalent and partially ionic. The difference in electronegativity determines the % ionic character.

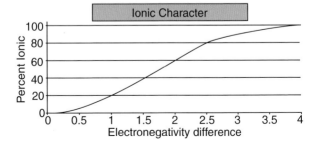

Ionic Character

7 Mixtures & Solutions

■ **Physical Combinations:** Solute (solid or liquid material present in the lesser amount) dissolves in the solvent (liquid or material present in the lesser-larger amount).

■ **Major Factor Promoting Mixing of Fluids:** A mixture is more disordered than separate phases.

■ **Variable Factor:** Interaction between molecules, general guide, "like-dissolves-like"
   ◆ polar solvents mix, they are miscible
   ◆ nonpolar solvents mix, they are miscible
   ◆ Polar and nonpolar solvents are usually immiscible, forming separate liquid layers.

■ **Solution Units:**
   ◆ **M: Molarity** - moles of solute dissolved in 1 liter of solution.
   ◆ **m: Molality** - moles of solute per kg of solvent.
   ◆ **x: Mole fraction** - mole of solute divided by total moles in the solution.

■ **Colligative Properties:** Depend only on the number of solute particles and the identity of the solvent. Ionic materials dissociate.
   ◆ vapor pressure lowering:
   ◆ $\Delta P = x_{solute}\ P^0$ (solvent vapor pressure)

■ **Freezing Point Depression:**
   ◆ $\Delta T = -m_{solute}\ k_{fp}$ (constant depends on solvent)

■ **Boiling Point Elevation:**
   ◆ $\Delta T = m_{solute}\ k_{bp}$ (constant depends on solvent)

■ **Osmotic Pressure, Π:**
   ◆ $\Pi = M_{solute}\ RT$
   ◆ M is the molarity of the solution.
   ◆ R is the Ideal Gas constant
   ◆ Osmotic pressure accounts for turgor pressure in plants and shape in animal cells.

8 Gases

- ■ **P: Pressure** is the force/area exerted on the container walls (in atm, mm Hg, Pa)
- ■ **V: volume** of the gas sample (in liters)
- ■ **T: Temperature** (in Kelvin)
- ■ **n: moles** of gas particles
- ■ **R: Ideal Gas Constant STP** (Standard Temperature and Pressure): 1 atm; 273.15 K, one mole of **Ideal Gas** occupies 22.414 liter
- ■ **Boyle's Law:** P x V is a constant for fixed T; $P \propto 1/V$

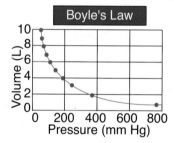

- ■ **Charles' Law:** $V \propto T$ fixed P

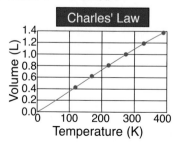

- ■ **Avogadro's Law:** $V \propto n$ for fixed T and P

23

■ **Ideal Gas Law: PV = nRT**
Model relationship for describing properties of a gas sample. Non-interacting massless particles. reliable for low P, high T.

■ **Velocity & Energy:** The kinetic energy of 1 mole of Ideal Gas is $\frac{3}{2}RT$; the velocity (rms: root mean square) for a gas with molar mass "M" at a given T is

$$v_{rms} = \sqrt{\frac{3RT}{M}}$$

The average speed of a typical gas molecule ranges from 400-1800 m/sec at STP.

■ **Graham's Law:** Effusion is gas expanding into a vacuum. The rate of effusion is proportional to

$$\frac{1}{\sqrt{M}}$$

Graham's Law also describes the diffusion of gases in mixtures of gases

■ **Gas Mixtures: Dalton's Law of Partial Pressure**
$P_{total} = P_1 + P_2 + . . P_i$; total pressure equals the sum of gas partial pressures, P_i ; V and T are constant.
$P_i = x_i P_{total}$ (x_i = mole fraction of gas "i").

■ **Real gases violate the Ideal Gas Law**
◆ PVT data does not fit Ideal predictions
◆ Gases liquefy at low T and high P
◆ Molecules have rotational and vibrational energy

The van der Waals equation improves the Ideal Gas model; adds terms for molecular volume and inter molecular attractions.

Bonding

Formal Bonding Models

■ **Lewis Structure:** Visual representation of a molecule with the valence electrons (dots) allocated as bonding-pairs and non-bonding lone-pairs.

$$H{:}H \qquad {:}\ddot{C}l{:}\ddot{C}l{:} \qquad {:}N{:::}N{:} \qquad \ddot{O}{::}\ddot{O}$$

$$H{:}\ddot{C}l{:} \qquad \ddot{O}{::}C{::}\ddot{O} \qquad H{:}\ddot{O}{:}H$$

■ **Octet Rule:** Atoms surrounded by 8 electrons, forming 4 bonds (except H which will have 2 electrons, and atoms with d-orbitals which can form 6 bonds, e.g. SF_6). The **"closed shell"** rule

■ **Multiple Bonds:** Bonded atoms may share 1, 2 or 3 electron-pairs.

■ **Bond Order:** The number of bonds for an atom divided by the number of neighbors. Single bond (order 1.0), double (2.0), triple (3.0), may be fractional. For a given bond-type, bond length decreases with increasing bond order.

■ **Resonance:** If different Lewis structures are possible, their average is the more accurate view of the bonding. Resonance delocalizes electrons and stabilizes energy.

For example, Ozone (O_3) has two resonance structures:

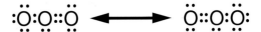

■ **Formal Charge:** Denotes that an atom has gained or lost valence electrons, relative to the free atom.
Calculate as Follows: (number of valence electrons in free atom) - ½ (number bonded electrons) - (number lone pair electrons).
Other Things Equal: structures with smaller formal charges are more stable, place negative charge on more electronegative atom.

Valence Bond Theory

■ A chemical bond forms by the overlap of singly-occupied valence atomic or hybrid orbitals. The resulting pair of electrons form the chemical bond.

Chemical Bonding & Quantum Mechanics

■ **Particle-Wave Duality:** Electrons and light exhibit wave and particle character. The wavelength (λ) of **light** describes the "color," related to the frequency (ν) and speed of **light** (c) by $\lambda\nu = c$

■ The energy of light is quantized in photons, $h\nu$ (h, Planck's Constant).

■ In simple form, the electron wave-property is described by the **deBroglie** model; the λ is related to its mass (m) and velocity (v) by $\lambda = h/mv$.

■ Electrons in atoms occupy discrete energy levels and are described with quantum numbers and wavelike **atomic orbitals (AO's)**.

■ Electrons in molecules occupy discrete energy levels and are described with wavelike **molecular orbitals (MO's)**.

■ The Aufbau principle guides the filling of the atomic and molecular energy levels. The complete description of the wave character is termed the **wavefunction Ψ**. The AO's give Ψ for an atom.

■ The MO's give the Ψ for a molecule.

10 Oxidation-Reduction Reactions

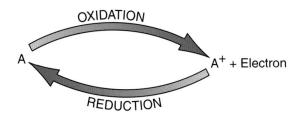

OXIDATION

A ⟶ A⁺ + Electron

REDUCTION

■ **Oxidation Number:** Element (0), ionic substance (charge of the ion), covalent compound (charge on atom if all valence electrons shift to the more electronegative atom)

■ **Redox Chemistry:** Electrons are exchanged in the reaction; oxidation numbers change.

◆ View reaction "A + B => AB" in two steps called **half-reactions:**

• first **oxidation:** (A loses electrons) $A => A^+ + e^-$

• then **reduction:** (B gains electrons) $B + e^- =>$ $B^-A^+ + B^-$ form the product AB

■ **Balancing Redox Reactions:** Two common approaches:

◆ **Half-Reaction Method:** Balance the reduction and oxidation "half-reactions", Combine, with electron flow balanced.

◆ **Oxidation-Number Method:** Identify changes in elements valence; balance electron exchange.

◆ For acidic: use H_3O^+ and H_2O to balance. For basic: use OH^- and H_2O.

Examples of Redox Reactions
- ◆ battery/galvanic:
  $Zn(s) + Cu^{2+}(aq) => Zn^{2+}(aq) + Cu(s)$
- ◆ electrolysis, $2 H_2O(l) => 2 H_2(g) + O_2(g)$
- ◆ corrosion: $2 Al(s) + 3O_2(g) => 2 Al_2O_3(s)$
- ◆ combustion: $C + O_2 => CO_2$

Electrochemical Cell: An external circuit connects two electrodes to facilitate the reaction.
- ◆ anode - site of oxidation
- ◆ cathode - site of reduction

Cell EMF, ε: electrical potential generated by the cell: $\varepsilon > 0$ for spontaneous process.

Galvanic (Voltaic) Cell: Spontaneous reaction produces a flow of current; used to make batteries.

Diagram for Zn, Cu cell:
$Zn(s) \mid Zn^{2+}(aq) \parallel Cu^{2+}(aq) \mid Cu(s)$

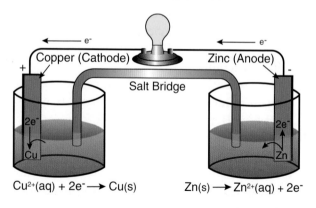

$Cu^{2+}(aq) + 2e^- \longrightarrow Cu(s)$ $Zn(s) \longrightarrow Zn^{2+}(aq) + 2e^-$

Electrolytic Cell: External current/voltage drives the reaction. Battery powered electrolytic process. For $\varepsilon < 0$, reverse is spontaneous.

■ $\Delta G = - n \, \mathcal{F} \varepsilon$ ($\mathcal{F}$, Faraday constant, "n" moles of electrons); reverse reaction, change the sign of the potential.

■ **Standard Potential:** $\varepsilon^0 = \varepsilon^0_{anode} + \varepsilon^0_{cathode}$

■ **Reduction Potentials:** Standard tabulation of electrode half-reactions (written as reduction).

◆ Reference potential: H_2 electrode 0.00 V

◆ Nonmetals: $X_2 + 2 \, e^- \Rightarrow 2 \, X^-$
A larger, positive number is evidence of a more reactive material.

F_2	2.87	Most reactive
Cl_2	1.36	
Br_2	1.09	
I_2	0.54	Least reactive

◆ Metals: $M^{x+}(aq) + x \, e^- \Rightarrow$ Metal
The more positive value is evidence of a less reactive metal.

Ag(I)	+0.80	Least reactive
Cu(II)	+0.34	
Pb(II)	-0.13	
Ni(II)	-0.26	
Fe(II)	-0.45	
Zn	-0.76	
Al	-1.66	
Mg	-2.37	
Na	-2.71	
Li	-3.04	Most reactive

Inorganic Salts

General Guidelines for Aqueous Solubility

	acetate nitrate perchlorate chlorate	chloride bromide iodide	fluoride	sulfate	carbonate sulfide phosphate hydroxide chromate	oxide
alkali metals, ammonium		Soluble				form hydroxide
Ca, Mg			insol	s	insol	insol
Sr, Ba			insol	insol	insol	insol
Fe, Cu, Zn			insol	s	insol	insol
Pb(II)		insol	insol	insol	insol	insol
Hg(I)		insol	-	insol	insol	insol
Ag		insol	s	s	insol	insol

Flame Test – Characteristic Ion Colors in a Flame

violet: potassium, rubidium, cesium

blue: copper (azure); lead, arsenic, selenium

green: copper (emerald), barium (yellowish), zinc (whitish)

yellow: sodium

red: lithium (carmine), strontium (scarlet), calcium (yellowish red)

12 Acid-Base Reactions

Water

■ Self-ionization of water: $2H_2O \leftrightarrow OH^- + H_3O^+$
  $K_w = [OH^-][H_3O^+] = 1 \times 10^{-14}$ at 25°C

■ For a neutral aqueous solution
  $[OH^-] = [H_3O^+] = 1 \times 10^{-7}$

■ $pH = -\log_{10}[H_3O^+]$: a measure of acidic strength:
  - **Neutral solutions, pH = 7**
  - **Acidic, pH < 7**
  - **Basic, pH > 7**

> **SAMPLE:**
> a 0.01 M solution of H_3O^+ has a pH of 2

■ pOH ($pOH = -\log_{10}[OH^-]$) can be used for basic solutions. $pH + pOH = 14$

Acids

■ Acids (HA): $HA + H_2O \iff A^- + H_3O^+$

A^- is the conjugate base of the acid HA

■ **Strong acid,** total dissociation:
HCl, HBr, HI, $HClO_4$, H_2SO_4 and HNO_3.

■ **Acid equilibrium** is described by
$$K_a = \frac{[A^-][H_3O^+]}{[HA]}$$

■ **Weak acids** have $K_a \ll 1$
$pK_a = -\log_{10}(K_a)$

■ **Common weak acids (pK_a):** acetic (4.19); HF (4.15); nitrous (3.35); carbonic (6.37)

■ **Lewis acid** is an electron pair acceptor.

■ **Polyprotic acid** is a compound with more than one ionizable proton (eg. H_2SO_4, H_3PO_4).

Bases

■ Bases (B):

$MOH \leftrightarrow OH^- + M^+$

or $B + H_2O \leftrightarrow HB^+ + OH^-$

HB⁺ is the conjugate acid of the base

■ **Strong base,** complete dissociation: NaOH, KOH and $Ba(OH)_2$

■ **Base equilibrium** is described by

$$K_b = \frac{[OH^-][HB^+]}{[B]}$$

■ **Weak bases** have $K_b \ll 1$.

$pK_b = -\log_{10}(K_b)$.

■ **Common weak bases** (pK_b):

NH_3 (4.75); CN^- (4.70).

■ **Lewis base** is an electron-pair donor.

Applications of Acids & Bases

■ **Amphoteric Substance:** A material which can react as an acid or a base

■ **Hydrolysis:** Water + aqueous ion produces an acidic or basic solution by forming OH^- or H_3O^+:

$A^- + H_2O \leftrightarrow HA + OH^-$ (basic, eg. F^- or acetate)

$HX^+ + H_2O \leftrightarrow X + H_3O^+$ (acidic, eg. NH_4^+)

■ **Buffer:** A solution of weak acid and a salt of its conjugate base, or a solution of a weak base and a salt of its conjugate acid. The mixture maintains "constant" pH, **Henderson-Hasselbalch** equation for an acid/salt buffer:

$$pH = pK_a + \log_{10} \frac{[salt]}{[acid]}$$

■ **Acid-Base Titration:** React a known amount of acid with a basic solution of unknown concentration. At the equivalence point:

◆ moles of acid = the moles of base.
◆ For strong acid-strong base titration pH = 7
◆ For weak acid-strong base titration pH >7
◆ For weak base-strong acid titration pH <7

SAMPLE:

Reactions: *Acid-rain:* Sulfur and nitrogen oxides react with water to give acids. Strong acid/bases react with metals, produce H_2 and salt. Carbonates are decomposed by acid. Copper + nitric acid produces nitrogen oxides.

13 Equilibrium

■ **Reactions Going to Completion:** All reactants are converted to products. ($\rightarrow$)

■ **Equilibrium:** The reaction reaches a steady-state of forward and reverse reactions. Products and reactants coexist. ($\leftrightarrow$)

■ For: $aA \leftrightarrow bB$: The equilibrium concentrations of reagents, $[A]_{eq}$ and $[B]_{eq}$, are constrained by the relationship:

$$K_{eq} = [B]_{eq}^b / [A]_{eq}^a$$

K_{eq} is a constant, characteristic of the reaction at a given temperature.

■ **Solubility Product, K_{sp}:** Defines the equilibrium between a salt and its aqueous ions

◆ For AX_2, the equilibrium is:

$$AX_2(s) \leftrightarrow A^{2+} (aq) + 2X^- (aq)$$

◆ $K_{sp} = [A^{2+}][X^-]^2$

◆ K_{sp} is small for a low solubility salt.

■ **LeChatelier's Principle:** The equilibrium shifts in response to changes in temperature, pressure or reagent concentration.
   ◆ A ↔ B, removing B or adding A shifts equilibrium towards the product.
   ◆ For increases in pressure, the equilibrium shifts to lower the total pressure (increasing the pressure raises the concentration). Most relevant for gas phase reactions.

■ **Exothermic** reaction produces heat:
   "A ↔ B + heat"
   ◆ lowering the temperature removes heat and shifts equilibrium towards the product.
   ◆ Raising the temperature has the opposite effect.

■ **Endothermic** reaction absorbs heat:
   "Heat + A ↔ B"
   ◆ Raising the temperature adds heat and shifts equilibrium towards the product.
   ◆ Lowering the temperature has the opposite effect.

14 Thermodynamics

The study of the heat and work associated with **a physical or chemical process.**

Types of Processes
- ◆ **Reversible,** the system is in a state of equilibrium.
- ◆ **Spontaneous** (irreversible), the system is moving towards a state of equilibrium.

Laws of Thermodynamics
- ◆ *First Law* - Conservation of energy, (U): The heat (q) and work (w) associated with a process are interrelated: $\Delta U = q+w$.
- ◆ Any change in the energy of the system must correspond to the interchange of heat or work with an external surrounding.
- ◆ *Second Law* - **Entropy**, S, is conserved for a reversible process. The disorder of the system and thermal surroundings must increase for a spontaneous process.
- ◆ *Third Law* - **Entropy** is zero for an ideal crystal at T=0 K. The system is in its lowest possible energy state and most ordered arrangement.

■ **Enthalpy (H)**: ΔH is the heat absorbed or produced by a process under conditions of constant pressure (normal lab conditions).

  ◆ Is heat released or absorbed?

    $\Delta H < 0$ for an exothermic reaction or process heat is released to the surroundings.

    $\Delta H > 0$ for endothermic reaction or process heat is absorbed from the surroundings.

  ◆ **Enthalpies of Formation**: ΔH_f^0

    The ΔH for the synthesis of the compound from the standard elemental forms at 25°C.

    These quantities can be either positive or negative

  ◆ **Enthalpy change for a process or reaction**:

    ΔH = (sum of product ΔH_f^0) - (sum of reactant ΔH_f^0).

Entropy (S): Thermodynamic disorder:
ΔS is the change in order in a system

SAMPLE:
For a phase transition from solid $\rightarrow$ liquid or liquid $\rightarrow$ gas, ΔS is positive
(in each case the product has more random motion than the reactant)

◆ **Standard entropy, S^0**: The entropy of a compound at 25°C relative to 0 K.

NOTE:
These quantities are always positive

◆ **Entropy change for a process or reaction**:
ΔS = **(sum of product S^0) - (sum of reactant S^0)**

■ **Gibbs Free Energy** (G): $\Delta G = \Delta H - T\Delta S$.

ΔG: The capacity of the system to perform work

ΔG=0 for equilibrium

ΔG<0 for spontaneous

ΔG>0 the reverse process is spontaneous

◆ **Free energy of formation, ΔG_f^0:** The ΔG for the synthesis of the compound from standard elemental forms at 25°C.

◆ **Free energy change for a process or reaction:**

ΔG = (sum of product ΔG_f^0)-(sum of reactant ΔG_f^0)

◆ **Free energy and equilibrium**

The equilibrium constant, $\mathbf{K_{eq}}$, and **ΔG** are related by the equation:

$$\Delta G = -RT \ln(\mathbf{K_{eq}})$$

NOTE:
Large negative **ΔG** corresponds to a large equilibrium constant-the reaction is spontaneous

■ **Rates of Chemical Process**
For a generic reaction: $aA + bB \Rightarrow cC + dD$, the rate usually depends on [A] and [B]

■ **The Rate Law:** the mathematical equation that describes the reaction rate as a function of reagent concentrations

■ **First Order Rate Law:** Rate = $k_1[A]$

◆ If you double the concentration of A, you double the rate of the reaction

◆ A graph of "ln[A] vs. time" is linear, the slope is the rate constant k_1

◆ **half-life,** $t_{1/2}$: The time required for the concentration to decrease by a factor of 2

$$t_{1/2} = 0.693/k_1$$

SAMPLE:
Radioactive decay of uranium-235 is a first order process

■ **Simple Second Order Rate Law:** Rate $= k_2[A]^2$

◆ If you double the concentration of A, you quadruple the rate of the reaction

◆ A graph of "$1/[A]$ vs. time" is linear, the slope is the rate constant k_2

◆ The reaction half-life changes during the reaction.

■ **Zero Order:** Rate $= k_0[A]^0 = k_0$
The rate is independent of [A]

NOTE:
$[A]^0 = 1$

■ **Temperature-Dependence** of rate constants
Arrhenius Law:

$$k = A \, e^{-E_a/RT}$$

◆ E_a, is the activation energy: the energy barrier
   that separates reactants and products

◆ *a graph of "ln(k) vs. 1/T"* is linear, the slope is
   $-E_a/R$, and the intercept is *ln(A)*

■ **Kinetics & Thermodynamics**

For a simple equilibrium, A ↔ B, the equilibrium
constant is related to the forward rate constant, k_1,
and reverse rate constant, k_{-1}, by the following:

$$K_{eq} = k_1/k_{-1}$$

Hints for Balancing Equations

Find whole number coefficients which give the same amount of each element on each side of the equation.

- Identify each element involved in the reaction.
- Change coefficients only, not the formulas.
- Apply coefficients to each atom in a polyatomic ion.
- Determine the net charge for each side of the equation: must be balanced in the final equation.
- Start with the element appearing once on each side. Next focus on the more complex compounds.
- If an element appears in a pure form, leave it to the last step.
- It may help to use fractions to balance, then convert to integer coefficients after all elements are balanced.
- Final step, make sure coefficients are the smallest whole numbers.

Remember: Always check your work! Make sure that the same number of each type of atom and the same total charge are on each side of the equation.

Basic Skills

Calculator Survival

◆ Become familiar with your calculator before the exam; make sure you can multiply, divide, add, subtract and use all needed functions.

◆ Calculators never make mistakes; they take your input (intended & accidental) and give an answer.

◆ Look at the answer; does it make sense?

◆ Do a quick estimate to check your work.

SAMPLE:

$\dfrac{4.34 \times 7.68}{1.05 \times 9.8}$ is roughly $\dfrac{4 \times 8}{1 \times 10} = 3.2$

the actual answer is **3.24**

How to Do Word Problems

◆ Read and evaluate the question before you start plugging numbers into the calculator.

◆ Identify the variables, constants and equations.

◆ Write out units of the variables and constants.

◆ Work out the unit before the number.

◆ You may have extra information, or you may need to obtain constants from your text.

▨ How to Work with Units

All numerical data has units. In chemistry we use metric "SI" units.

⚠ **Pitfall:** If the unit is wrong, the answer is wrong!

◆ Unit prefixes: denote powers of "10"

tera	T	10^{12}	giga	G	10^9
mega	M	10^6	kilo	k	10^3
deci	d	10^{-1}	centi	c	10^{-2}
milli	m	10^{-3}	micro	μ	10^{-6}
nano	n	10^{-9}	pico	p	10^{-12}
femto	f	10^{-15}			

◆ **Fundamental Units**
  • **Mass:** kilogram (kg)
  • **Length:** meter (m)
  • **Temperature:** Kelvin (K)
  • **Time:** second (s)
  • **Amount of a substance:** mole
  • **Electric charge:** coulomb, (C)

◆ **Derived Units**
  • **Area:** length squared, m^2
  • **Volume:** length cubed, m^3; 1 liter (L) = 1 dm^3
  • **Density:** mass/volume; common unit kg/m^3
  • **Speed:** distance/time; common unit m/s
  • **Electric current:** ampere (A) = 1 C/s
  • **Force:** Newton (N) = 1 $kg\ m/s^2$
  • **Energy:** Joule (J) = 1 $kg\ m^2/s^2$
  • **Pressure:** Pascal (Pa) = 1 $kg/(m\ s^2)$

◆ Fundamental Constants

Mass	in amu (g/mole)	in kg
electron:	5.486 x 10⁻⁴	9.10939×10^{-31}
proton:	1.007276	1.67262×10^{-27}
neutron:	1.008664	1.67493×10^{-27}

Electronic charge: 1.6022×10^{-19} C
Avogadro's number, N_A: 6.02214×10^{23}
Ideal gas constant, R:
R, for gas calculation: 0.082 L atm/(K mol)
R, for energy calculation: 8.3145 J/(K mol)

Faraday Constant, $\mathcal{F}$:   96,485 C/mol
Planck's Constant, h: 6.626×10^{-34} J s
Speed of Light, c: 2.9979×10^8 m/s

▧ How to Convert Data
   ◆ The unit & numerical value are **changed** using a
      conversion factor or equation.

▧ How to Use Equations for Data Conversion

SAMPLE:
Convert a temperature of 45°C to °F and
Kelvin:
 Given: K = °C + 273.15
 Calculate temperature in Kelvin
 K = 45°C + 273.15 = 318.15 K

 Given: °F = 9/5 °C + 32
 Calculate temperature in Fahrenheit
 °F = 9/5 x 45 °C + 32= 81 + 32 = 113 °F

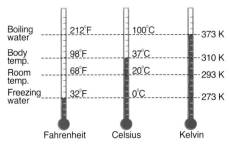

	Fahrenheit	Celsius	Kelvin
Boiling water	212°F	100°C	373 K
Body temp.	98°F	37°C	310 K
Room temp.	68°F	20°C	293 K
Freezing water	32°F	0°C	273 K

How to Use a Conversion Factor

SAMPLE:
"1 hour = 60 minutes" gives 2 conversion factors:
◆ **Multiply by** "1 hour/60 minutes" to convert minutes to hours.
◆ **Multiply by** "60 minutes/1 hour" to convert hours to minutes.

SAMPLE:
Determine the number of hours in 45 minutes.
◆ The conversion factor is "1hour/60 minutes."
◆ **Calculate:** Time = 45 minutes x (1 hour/60 minutes) = 0.75 hour (minutes cancel)

◆ **Common conversion factors:**
1 calorie = 4.184 J   1 kg = 2.2 lb 1 m = 1.1 yd
1 qt = 0.9464 L   1 angstrom (Å) = 1 x 10^{-10} m
1 atm = 1.01325 x 10^5 Pa   1 atm = 760 mm Hg

⚠ **Pitfall:** Equations and conversion factors have units. Beware of inverting conversion factors.

Math Review

How to Work with Algebraic Equations
Give equal treatment to each side.

◆ **Add or subtract:**

SAMPLE:
Given, $x = y$, then, $4 + x = 4 + y$

◆ **Multiply or divide:**

SAMPLE:
Given, $x = y$, then, $4x = 4y$ and $x/5 = y/5$

Given: $a = b+4$, then $\dfrac{a}{x-2} = \dfrac{b+4}{x-2}$

⚠ **Pitfall:** Dividing by zero is not allowed.

How to Work with Scientific Notation
The exponent gives the power of 10

SAMPLE:
$0.00045 = 4.5 \times 10^{-4}$ $1345 = 1.345 \times 10^3$

Chemical applications: Molecular diameters are 10^{-10} m; 1 liter of water contains 1×10^{26} atoms.

How to Use Logarithms & Exponents
◆ **Common logarithm,** $\log_{10}$: base "10".
Denotes number or function in powers of 10

SAMPLE:
Given, $y = 10^6$, then $\log_{10} y = 6$.

◆ **Natural logarithm,** ln: base "e" (e = 2.71828)
Denotes number or function in powers of "e"

SAMPLE:
Given, $z = e^5$, then $\ln z = 5$

◆ **Working with logs**
Products: $\log (xy) = \log x + \log y$
Powers: $\log (x^n) = n \log (x)$
Multiplication: Add exponents from each term:

SAMPLE:
$10^5 \times 10^3 = 10^{5+3} = 10^8$

Division: Subtract denominator exponents from numerator exponents:

SAMPLE:
$10^5/10^3 = 10^{5-3} = 10^2$

Square root: $\sqrt{a} = a^{1/2}$
Inverse: $\frac{1}{x} = x^{-1}$

⚠ **Pitfall:** Your calculator has separate keys for lnx (base e), logx (base 10), 10^x and e^x.

◆ **Chemical applications of logs:**
pH of acid and base

▨ How to Calculate Roots of a Polynomial

An equation of the form: $ax^2 + bx + c = 0$, has 2 solutions or roots, given by the **quadratic** formula:

$$\frac{-b + \sqrt{b^2 - 4ac}}{2a} \quad \text{Root 1}$$

$$\frac{-b - \sqrt{b^2 - 4ac}}{2a} \quad \text{Root 2}$$

SAMPLE:

Determine the roots for the equation:

$3x^2 + 4x + 1 = 0$

Given: $a = 3$, $b = 4$, $c = 1$

Calculate: Root 1 $= \dfrac{-4 + \sqrt{4^2 - 4 \times 3 \times 1}}{2 \times 3}$

Root 1 $= -\frac{1}{3}$

Calculate: Root 2 $= \dfrac{-4 - \sqrt{4^2 - 4 \times 3 \times 1}}{2 \times 3}$

Root 2 $= -1$

⚠ **Pitfall:** Beware round-off error. Substitute the roots into the equation to verify results.

◆ **Chemical applications:** Weak acids, weak bases, buffers, chemical equilibrium

How to Determine the Equation of a Line

Linear Equation: $y = mx + b$

m: **slope** of the line; $m = \Delta y/\Delta x$

b: **y intercept,** the line crosses the y-axis at "b"

$$b = y_i - m\, x_i$$

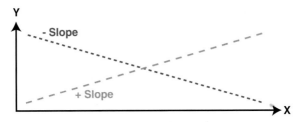

SAMPLE:

Determine the equation of a line using (x,y) data.

Given: x: -2   -1   0   1   2   3
            y: -2   1   4   7   10   13

Calculate: Slope $= m = \Delta y/\Delta x = \dfrac{13-(-2)}{3-(-2)} = \dfrac{15}{5} = 3$

Calculate: y-intercept $= b = y_i - m\, x_i = (-2) - 3 \times (-2) = -2 + 6 = 4$

The equation of the line is $y = 3x + 4$

◆ **Chemical applications:** Gas law calculations; Beer's Law; analyzing reaction-rate data

Statistics

How to Calculate Experimental Error

◆ **Accuracy:** The agreement between experimental data and a known value.

◆ **Error:** A measure of accuracy:

Error = (experiment value - known value)

Relative error = $\frac{\text{Error}}{\text{known value}}$

% error = relative error x 100%

SAMPLE:

A student finds the mass of an object to be 5.51g; the correct value is 5.80g. What is the % error?

Given:

experimental value = 5.51g;
known value = 5.80g.

Calculate:

error = (exp. value - known value) = 5.51- 5.80g = -0.29g

Calculate:

relative error = error/known value = -0.29g/5.8g = - 0.050

Calculate:

% error = relative error x 100% = - 0.05 x 100% = - 5.0%

In this case, we undershoot the known value by 5%.

◆ **Precision:** The degree to which a set of experimental values agree with each other.

⚠ **Pitfall:** A set of data can be precise, but have a large experimental error.

Calculating Mean & Deviation

For a set of numbers, $\{x_1, x_2, x_3 \dots x_j\}$:

◆ The mean, or average value, is the sum of all "x" divided by "j", the number of entries.

◆ The deviation, Δ_i, for each x_i

$$\Delta_i = x_i - \text{mean}$$

Δ_i can be positive or negative.

SAMPLE:

Determine the mean and deviations for the following data: $\{25, 28, 31, 35, 43, 48\}$

$$\text{Mean} = \frac{25+28+31+35+43+48}{6} = 35$$

x_i	25	28	31	35	43	48
Δ_i	25-35	28-35	31-35	35-35	43-35	48-35
Δ_i	-10	-7	-4	0	+8	+13

Note: Sum of $\Delta_i = 0$

How to Identify Significant Figures (sig. figs.)
- ◆ Record all certain digits and one uncertain or estimated digit for a measurement.

⚠ **Pitfall:** A calculator often displays extra digits in a calculation.

- ◆ For a multi-step problem, keep 1 or 2 extra digits; then round off the final answer

Rules for the number of sig. figs. in the final answer:
- ◆ For addition or subtraction: use the least number of **decimal places** found in the data.

SAMPLE:
$10.102 + 5.03 = 15.13$ *truncate to 2 **decimals***

- ◆ For multiplication or division: the final answer should have same number of **sig. figs.** as the entry with the fewest sig. figs.

SAMPLE:
$5.46200 \times 4.00 = 21.8$ *truncate to 3 **sig. figs.***

- ◆ **Rounding-off data:**
  Round up if the last uncertain digit is 6,7,8,9
  Round down if it is 0,1,2,3,4
  If it is a 5: the arbitrary convention is to round up if the last certain digit is odd, round down if it is even.

SAMPLE:
$0.085 \rightarrow 0.08$ $0.035 \rightarrow 0.04$
$0.453 \rightarrow 0.45$ $0.248 \rightarrow 0.25$

Properties of Atoms

■ **Atomic Number:** Z = number of protons in the nucleus

■ **Atomic Mass Number:** A = number of protons + number of neutrons in the nucleus

■ **A and Z are Integers.** The actual particle mass is given in kg or amu (g/mole). The actual mass is close in value to A.

■ **How to Calculate Nuclear Binding Energy**
   ◆ The nuclear mass does not equal the sum of proton and neutron masses.
   ◆ The mass difference (Δm, kg/mole) is due to the nuclear binding energy (ΔE, J):
$$\Delta E = \Delta m \, c^2$$

SAMPLE:
Calculate the binding energy for He-4.
Given: He-4 has 2 protons and 2 neutrons
Given: Proton particle mass: 2 x 1.00728 amu
Given: Neutron particle mass: 2 x 1.00866 amu

Total particle mass = Sum of proton and neutron particle mass = 4.03188 amu

Actual mass of He-4 nucleus: 4.00150 amu

Δm = Particle mass - He-4 mass = 0.03038 amu = 0.03038 g/mole
Convert to kg: $\Delta m = 3.038 \times 10^{-5}$ kg/mole

Calculate the nuclear binding energy:
$\Delta E = \Delta m \, c^2$
$= 3.038 \times 10^{-5}$ kg/mole $\times (3.00 \times 10^8$ m/s$)^2$
$= 2.73 \times 10^{12}$ J/mole

⚠ **Pitfall:** Watch units of mass; you can work in kg/particle or amu (g/mole). Note: J = kg m²/s²

Calculating Atomic Weight

For an element with two isotopes, denoted a and b:
Atomic wt = mass$_a$ x fract$_a$ + mass$_b$ x fract$_b$

SAMPLE:
Calculate the atomic weight for chlorine.
Given: Chlorine has two isotopes

	Isotope mass	fractional abundance
Cl-35	34.968852	0.7577
Cl-37	36.965303	0.2423

Calculate :
Atomic wt = mass$_a$ x fract$_a$ + mass$_b$ x fract$_b$
$= 34.968853 \times 0.7577 + 36.965303 \times 0.2423$
$= 26.496 + 8.9566$
$= 35.45$ amu (4 sig. figs.)

⚠ **Pitfall:** Use the actual mass of the nucleus, not the mass number.

How to Calculate Properties of Electromagnetic Radiation

Light waves are characterized by **wavelength** (λ, in m), **frequency** (ν, in Hertz, s^{-1}) and the speed of light (c, in m/s). The energy is carried in **photons.**

$$\lambda\nu = c$$
Energy of a photon = hν

SAMPLE:
Calculate the ν and energy for light with a λ of 500 nm
Given: λ = 500 nm and 1 nm = 1 x 10^{-9} m
Convert λ to m: λ = 500 nm = 5.00 x 10^{-7} m

Calculate:
ν = c / λ = 3.00 x 10^8 m/s / 5.00 x 10^{-7} m
= 6.00 x 10^{14} s^{-1} (Hertz)

Calculate:
energy = hν
= 6.626 x 10^{-34} J s x 6.00 x 10^{14} s^{-1}
= 3.98 x 10^{-19} J
(the "s" cancels)

⚠ **Pitfall:** The unit on λ should match the unit of c.

Chemical Formulas & the Mole

The **formula** and name denote elements and relative composition in the compound. A balanced equation conserves atoms and moles of each element.

◆ One mole is Avogadro's number, $N_A = 6.023 \times 10^{23}$, of atoms or molecules.

◆ The atomic weight, the mass in grams, of one mole of the atomic element as found in nature, is often given on the Periodic Table, along with the atomic number and element symbol.

◆ The molar mass of a compound is the mass, in grams, of 1 mole of the substance.

■ How to Calculate Molar Mass from the Formula

Given the atomic weight (at.wt.) of each element in the compound and the formula coefficients.

The molar mass is given by the sum of each element's atomic weight multiplied by the formula coefficient.

SAMPLE:

simple case: $MgCl_2$

Given:
Mg at.wt.:   24.305 g/mole; coefficient = 1
Cl at.wt.:   35.453 g/mole; coefficient = 2

Calculate:
Molar mass = 1 x Mg at.wt.+2 x Cl at.wt.
= 1 x 24.305 + 2 x 35.453
= 95.211 g/mole

SAMPLE:
complex case: $Mg(NO_3)_2$ x $2H_2O$

Given:
Mg at.wt.: 24.305 g/mole; coefficient = 1
N at.wt.: 14.007 g/mole; coefficient = 2
O at.wt.: 15.9994 g/mole; coefficient = 8
H at.wt.: 1.008 g/mole; coefficient = 4

Calculate:
Molar mass = 1 x Mg at.wt. + 2 x N at.wt.
+ 8 x O at.wt. + 4 x H at.wt
= 24.305 + 2 x 14.007 + 8 x 15.9994 + 4 x 1.008
= 24.305 + 28.014 + 127.9952 + 4.032
= 184.35 g/mole

⚠ **Pitfall:** It is easy to miscount the atoms in polyatomic ions or waters of hydration.

How to Calculate Elemental Percent Composition
The portion of the mass coming from each element in the compound; the %-compositions sum to 100%.
◆ **Given:** Chemical formula, compound molar mass and atomic weights for each element.
◆ **Step 1:** Sum the formula coefficients to determine the number of atoms of each element.
◆ **Step 2:** The mass of each element = at.wt. x number of atoms of that element.
◆ **Calculate:** % comp for element A

$$\frac{100\% \times \text{mass of A}}{\text{compound molar mass}}$$

SAMPLE:
Determine the elemental % comp of Mg and Cl in $MgCl_2$.

Given:
$MgCl_2$ molar mass =   95.21 g/mole
Mg at.wt. =   24.305 g/mole
Cl at.wt. =   35.453 g/mole

Step 1: $MgCl_2$ has 1 atom of Mg and 2 atoms of Cl

Step 2:
Mass of Mg = 24.305 g/mole x 1 = 24.305 g/mole
Mass of Cl = 35.453 g/mole x 2 = 70.906 g/mole

Calculate:
Mg % comp = 100% x 24.305 / 95.21= 25.53%
Cl % comp = 100% x 70.906 / 95.21 = 74.47 %

Sum of % comp = 25.53% + 74.47%= 100.00%

⚠ **Pitfall:** In a complex formula, the same element may exist in different ions.

How to Calculate the Number of Molecules

◆ **Given:** x, the mass of the sample, and the molar mass of the material

◆ **Calculate:**

$$\text{Number of moles} = \frac{x\,(\text{in g})}{\text{molar mass}\,(\text{g/mole})}$$

◆ **Calculate:** Number of molecules = number of moles x Avogadro's number (N_A)

SAMPLE:

Determine the number of moles and the number of water molecules in 5.00 grams of water vapor.

Given:
5.00 g of H_2O
molar mass = 18.015 g/mole

Calculate number of moles of water =

$$\frac{5.00\text{ g }H_2O}{18.015\text{g /Mole }H_2O} = 0.278 \text{ moles } H_2O$$

Calculate number of molecules
= 0.278 moles H_2O x 6.022 x 10^{23} molecules/mole
= 1.67 x 10^{23} H_2O molecules

How to Calculate Empirical Formulas

◆ **Given:** % composition and atomic weights, start by converting "%" to "grams of element"; assume you have 100 g of sample.

◆ **Calculate** Moles for each element = grams of element **divided by** atomic weight of the element

◆ **Calculate** The formula coefficient for each element equals the number of moles divided by the smaller number of moles calculated in the previous step.

◆ To determine the **empirical formula**, multiply each coefficient by a number to give a whole number coefficient.

SAMPLE:

Determine the empirical formula for a compound containing 75% C & 25 % H by mass.

Given the Atomic weights: C at.wt. = 12.011
H at.wt. = 1.008
Assume you have 75 g of C and 25 g of H
(total mass = 100 g)

Calculate the moles for each element:

$$\frac{75g\ C}{12.011g\ /\ Mole\ C} = 6.24 \text{ moles of C}$$

$$\frac{25gH}{1.008g\ /\ Mole\ H} = 24.8 \text{ moles of H}$$

Calculate the coefficient for each element:
C formula coefficient: 6.24/6.24 = 1.00
H formula coefficient: 24.8/6.24 = 3.97

The empirical formula is CH_4.

⚠ **Pitfall:** Due to experimental error, the calculated formula coefficients may not be integers.

■ How to Derive the Molecular Formula from the Empirical Formula

The molecular formula coefficients are obtained by multiplying each empirical formula coefficient by a **factor:** $\dfrac{\text{molecular molar mass}}{\text{empirical molar mass}}$

◆ **Given** the empirical formula, molecular molar mass, and atomic weights

◆ **Calculate the** empirical molar mass from the sum of atomic weights in the empirical formula

◆ **Calculate** the $\dfrac{\text{molecular molar mass}}{\text{empirical molar mass}}$ factor.

Determine the molecular formula by multiplying each empirical coefficient by this factor

---

SAMPLE:

Determine the molecular formula for a compound with molar mass 28 and empirical formula CH_2

Given the empirical formula is CH_2;
C atomic weight = 12.0; H atomic weight = 1.0

Calculate: the empirical molar mass
= 12 + 2 = 14
The molecular/empirical factor = 28/14 = 2
The **molecular** formula is C_2H_4
These coefficients are **twice** the **empirical** formula coefficients

Stoichiometry

How to Use Balanced Equations
◆ Verifying conservation of mass and moles

SAMPLE:
Balanced Equation
$2 Mg + O_2 \rightarrow 2 MgO$

2 Mg atoms and 2 O atoms on each side
2 moles of Mg and 2 moles of O on each side

Calculate the masses of the product and reactants:
Given a balanced equation and atomic weights:
$Mg = 24.31$; $O = 16.00$

Mass of Mg = 2 x Mg at.wt. = 48.62g of Mg
Mass of O_2 = 2 x O at.wt. = 32.00g of O_2
Mass of MgO = 2 x Mg at.wt.+2 x O at.wt.
= 80.62g of MgO

48.62g of Mg reacts with 32.00g of O_2 to produce 80.62g of MgO

A balanced equation conserves mass, 80.62g on each side, and conserves moles of each element

How to Balance an Equation

◆ **First:** Identify each element in the reaction.

◆ **Next:** Determine the net charge on each side; this must be balanced in the final equation.

◆ **Guidelines:**
  • Start with the element found in one compound on each side.

  • Identify compounds that must have the same coefficient.

  • If an element appears in a pure form, leave it to the last step.

  • You should always check your work!

⚠ **Pitfall:** Coefficients apply to each atom in a molecule. You change coefficients in the equation, **not** the formula subscripts.

SAMPLE:
Fill in the (?)
? CH_4 + ? O_2 → ? CO_2 + ? H_2O

First: Elements: C, H, O

Next: No charge to worry about.

1. **H is in CH_4 and H_2O; start with H**
2. **The H_2O coefficient must by twice the CH_4 coefficient to balance "H."**
   • 1 CH_4 + ? O_2 → ? CO_2 + 2 H_2O

3. **CH_4 and CO_2 must have the same coefficient**
   • 1 CH_4 + ? O_2 → 1 CO_2 + 2 H_2O

4. **Now determine the O_2 coefficient:**
   • 1 CH_4 + 2 O_2 → 1 CO_2 + 2 H_2O

5. **Check your work:**
   1 C, 4 H and 4 O on each side...
   the equation is balanced!

Balancing a Redox Equation Using the Half-Reaction Method

◆ Split into oxidation and reduction half-reactions.

◆ You may need to add H_2O and H^+ for acidic, or H_2O and OH^- for basic reaction conditions. This provides oxygen for the reaction

◆ Balance the half reactions separately, then combine them to balance the exchange of electrons.

⚠ **Pitfall:** Make sure you use H_2O, H^+ (for acidic solutions), or OH^- (for basic solutions) with the correct half reaction.

SAMPLE:

Balance the following for acidic solution:

- ? MnO_4^- + ? Fe^{2+} → ? Mn^{2+} + ? Fe^{3+}

1. In acidic solution: add H^+ to the left and H_2O to the right side:

- ? MnO_4^- + ? Fe^{2+} + ? H^+ → ? H_2O + ? Mn^{2+} + ? Fe^{3+}

2. Identify the half-reactions:

- Fe^{2+} → Fe^{3+} (oxidation)
- MnO_4^- + H^+ → Mn^{2+} + H_2O (reduction)

3. Add electrons to account for valence changes;

- Fe^{2+}→Fe^{3+} +1e^- Fe(II) to Fe(III)
- 5e^- + MnO_4^- + H^+→Mn^{2+} + H_2O Mn(VII) to Mn(II)

4. Balance each half-reaction:

Oxidation: Multiply by a factor of 5 to match electrons in reduction step:

- 5 Fe^{2+} → 5 Fe^{3+} + 5e^-

Charge: +10 on each side, balanced!

Reduction: Balance O, then H^+; check charge:

- 5e^- + MnO_4^- +8H^+ →Mn^{2+} + 4H_2O

Charge: +2 on each side, balanced!

5. Combine half-reactions to eliminate the 5e^-

- 5 Fe^{2+} + MnO_4^- + 8H^+ → Mn^{2+} + 4H_2O + 5Fe^{3+}

Check your work: 5 Fe, 1 Mn, 4 O and 8 H on each side, atoms are balanced!

Charge: +17 on each side, charge is balanced!

How to Calculate the Theoretical-Yield

The mass of a reactant is used to determine the mass of a product.

Given mass of reactant, balanced equation, molar masses of reactants and products

Calculate moles of reactant

$$= \frac{reactant\ mass}{reactant\ molar\ mass}$$

Calculate molar ratio

$$= \frac{product\ equation\ coefficient}{reactant\ equation\ coefficient}$$

Calculate moles of product
= moles of reactant x molar ratio

Calculate mass of product
= moles of product x product molar mass

⚠ **Pitfall:** If your balanced equation is wrong, your theoretical yield will usually be wrong.

SAMPLE:

Calculate the mass of Mg produced by burning 10.0g of Mg in excess oxygen.

Balanced equation: $2Mg + O_2 \rightarrow 2MgO$

Given 10.0 grams of Mg, Mg at.wt. = 24.305 g, MgO molar mass = 40.305 g

$$\text{Moles of Mg} = \frac{10.0g\ Mg}{24.305g\ /\ mol\ Mg}$$

= 0.411 moles of Mg

Molar ratio = 2/2 = 1

Calculate Moles of MgO

= 0.411 moles Mg x 1 mole MgO/1 mole Mg

= 0.411 moles of MgO

Calculate Mass of MgO =

0.411 moles MgO x 40.305 g MgO/mole MgO

= 16.6 g MgO

▧ Identifying the Limiting-Reagent for 2 Reactants

In a reaction with 2 reactants, the mass of product is constrained by the reactant in shortest supply, the **limiting reagent.**

◆ **Given:** balanced equation, mass of reactants, molar masses of reactants; specify number of reactants.

◆ **Calculate:** Moles of each reactant

$$= \frac{mass}{molar\ mass}$$

◆ **Calculate:** Ideal reactant molar ratio

$$= \frac{coefficient\ of\ reactant\ \#1}{coefficient\ of\ reactant\ \#2}$$

◆ **Calculate:** Actual reactant molar ratio

$$= \frac{moles\ of\ reactant\ \#1}{moles\ of\ reactant\ \#2}$$

◆ **Determine the limiting reagent:**
  • If actual reactant molar ratio ≤ ideal reactant molar ratio, then reactant #1 is the limiting reagent.
  • If actual reactant molar ratio > ideal reactant molar ratio, then reactant #2 is the limiting reagent.
  • Calculate the theoretical yield based on the mass of the **limiting reagent**.

  Hint: The reactant numbering is arbitrary, but you must stick with your choice for the entire calculation.

SAMPLE:
10.0g Mg reacts with 10.0g O_2

How much MgO is produced?
Balanced equation: $2Mg + O_2 \rightarrow 2MgO$

Given:
Reactant #1 is Mg, the molar mass = 24.305g
Reactant #2 is O_2, the molar mass = 32.00g

Calculate:
Moles of Mg = $\dfrac{10.0g\ Mg}{24.305g\ /\ mol\ Mg}$ = 0.411 mol Mg

Moles of O_2 = $\dfrac{10.0g\ O_2}{32.00g\ /\ mol\ O_2}$ = 0.3125 mol O_2

Calculate Ideal reactant molar ratio = 2/1 = 2

Calculate Actual reactant molar ratio
= moles Mg/ moles O_2 = 0.411/0.3125 = 1.31

Determine limiting reagent: 1.31 is less than 2.0, therefore Mg is the limiting reagent

Calculate the yield of MgO based on 10.0 grams of Mg (shown in the previous section)

⚠ **Pitfall:** Make sure you distinguish between the ideal and actual molar ratios.

Working with Gases
Simple model for gas behavior

$$PV = nRT$$
Ideal Gas Law

Pressure, P; common units: atm, Pa, bar or mm Hg; The R given below is for P in **atm.**
Volume, V; common units: liter (L), m
The R given below is for V in **L.**
Temperature, T; common units: Kelvin, ^oC or ^oF; always convert ^oC and ^oF to **Kelvin.**
Number of moles, n; moles = gas mass/gas molar-mass
Ideal Gas constant = R = 0.082 L atm/(mol K)
$$R = 0.082 \text{ L atm}^{-1} \text{ K}^{-1}$$

⚠ **Pitfall:** All data must fit the units of R.

■ **How to Calculate the Number of Moles of a Gas Sample**
  Given: Mass of gas (g), molar mass of gas
  Calculate: Moles = mass of sample / **molar mass**

SAMPLE:
Determine Number of moles in 5.0g of H_2 gas.
Given:
5.0g sample, H_2 molar mass = 2.016 g/mole

Calculate:
moles of H_2 = 5.0g H_2 x 1 mole H_2/2.016g H_2 = 2.48 mol H_2

If you are given density and volume, first calculate mass of the gas:
$$\text{mass (g)} = \rho(\text{g/L}) \times V(\text{L})$$

How to Use the Ideal Gas Law $P = \frac{nRT}{V}$

SAMPLE:
Calculate the pressure for 2.5 moles of Ar gas occupying 3.5 liters at 25°C.
Make required changes to variables:
$T(K) = 25°C + 273.15 = 298.15K$

Calculate: $P = \frac{nRT}{V}$

$$P = \frac{2.5mol \times 0.082\ L\ atm\ mol^{-1}\ K^{-1} \times 298.15K}{3.5\ L}$$

= 17.5 atm (Other units cancel)

How to Use Avogadro's Law $V \propto n$

V is proportional to number of moles, with constant T & P.
This is a **direct-proportionality** problem:
Given:
The number of moles changes by a factor of z.
Calculate: $V_{fin} = z \times V_{init}$

SAMPLE:
A 2.0 mole gas sample occupies 30.0L.
Determine the volume for 1.0 mole of the gas?

Given:
The number of moles, n, changes by a factor of ½

Calculate: $V_{fin} = \frac{1}{2} \times 30.0\ L = 15.0\ L$

How to Use Charles' Law

V and T are linearly proportional, with constant n & P

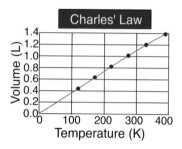

This is a **direct-proportionality** problem: $V \propto T$:
Given: T changes by a factor of z
Calculate: $V_{fin} = z \times V_{init}$

⚠ **Pitfall:** T must be in Kelvin; always convert °C to K.

SAMPLE:
A 3.5L sample of He gas is at 300 K; the temperature is raised to 900 K, what is the new volume?

Given: T increases three fold: 300 to 900 K

Calculate: $V_{fin} = 3 \times V_{init} = 3.0 \times 3.5 \text{ L} = 10.5 \text{ L}$

How to Use Boyle's Law

P & V are inversely proportional, with constant T & n.

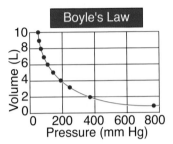

Boyle's Law

This is an **inverse-proportionality** problem:

◆ $P \propto 1/V$

   Given: V changes by a factor of z.

   Calculate: $P_{fin} = \dfrac{1}{Z} \times P_{init}$

◆ $V \propto 1/P$

   Given: P changes by a factor of z.

   Calculate: $V_{fin} = \dfrac{1}{Z} \times V_{init}$

SAMPLE:

The pressure of a 4.0L sample changes from 2.5 atm to 5.0 atm. What is the final volume?

Given: P changes two fold: 2.5 to 5.0 atm

Calculate: $V_{fin} = \frac{1}{2} \times V_{init} = \frac{1}{2} \times 4.0L = 2.0L$

How to Calculate the Speed of a Gas Molecule

$$v_{rms} = \sqrt{\frac{3RT}{M}}$$

NOTE:
R=8.314510 **J / (K mol)** =
8.314510 **kg m² / (s² K mol)**

SAMPLE:
Calculate the v_{rms} for He at 300 K
Given:
T = 300K
He, M = 4.00 g/mol = 4.00 x10⁻³ kg/mol

Calculate:

$$v_{rms} = \sqrt{\frac{3 \times 8.314 \text{ kg m}^2 \text{s}^{-2} \text{K}^{-1} \text{mol}^{-1} \times 300K}{4.00 \times 10^{-3} \text{ kg mol}^{-1}}}$$

$$= \sqrt{(1.87 \times 10^6 \text{ m}^2/\text{s}^2)} = 1370 \text{ m/s}$$

⚠ **Pitfall:** Watch the units on R and M; the final
unit is m/s. T must be in Kelvin.

How to Use Graham's Law of Effusion

The relative rate of effusion for molecules of mass

M_1 and M_2 $\dfrac{Rate_1}{Rate_2} = \sqrt{\dfrac{M_2}{M_1}}$

NOTE:
The inverse relationship between rate and mass

SAMPLE:
Determine relative rate of effusion for H_2 and CO_2

Given:
$M_1 = MH_2 = 2$ g/mol; $M_2 = MCO_2 = 44$ g/mol

Calculate:
Rate H_2/Rate $CO_2 = \sqrt{(44/2)} = 4.7$

H_2 diffuses 4.7 times as fast as CO_2

⚠ **Pitfall:** It is easy to invert the M_1/M_2; the smaller atom is always faster.

Solids & Liquids

Calculating Moles of Reagents

Determine the number of **moles** in "x-grams", or the **mass** needed to give a certain number of moles. Required data: the molar mass.

SAMPLE:

Calculate number of moles in 5.6 g of NaCl.

Given: NaCl molar mass = 58.44 g

Calculate Moles of NaCl

$$= \frac{5.6g \ NaCl}{58.44g \ / \ mol} = 0.096 \ moles \ NaCl$$

SAMPLE:

Calculate the mass of 0.25 moles of NaCl.

Given: NaCl molar mass = 58.44 g

Calculate Mass of NaCl

= 0.25 moles x (58.44g/mole)

= 14.61g NaCl

How to Calculate Mass of Liquids

Density (ρ), has units of g/mL

Pure reagent:

Use ρ & volume to determine the mass.

$$\textbf{vol x } \rho \textbf{ = mass}$$

SAMPLE:

Determine the mass of 30.0 mL of methanol

Given: ρ = 0.790 g/mL

Calculate the mass = vol x ρ

= 30.0mL x 0.790 g/mL = 23.7 g

▨ How to Calculate Moles of Reagents in Solution

Common units:

Molarity (M): moles of solute **per liter of solution**

molality (m), moles of solute **per kg of solvent**

Multiply the solution volume by the molarity to calculate **moles** of reagent.
Given: Solution molarity (M)
Calculate: Moles of reagent = vol x M

⚠ **Pitfall:** Volume should be in L. If you want to use "mL," denote M as "mmol/mL of solution"; 1 mmol = 0.001 mole.

SAMPLE:
Determine the moles and mass of NaCl in 25 mL of 2.35 M NaCl solution.

Given:
NaCl molarity = 2.35 M; molar mass = 58.44 g
Volume = 25 mL x 1 L/1000 mL = 0.025 L

Calculate:
NaCl moles = 2.35 mol/L x 0.025 L = 0.059 moles

Calculate:
NaCl mass = 0.059 moles x 58.44 g/mole = 3.45 g

How to Prepare Solutions

A solution is prepared by dissolving a known mass of solid in a specific amount of solvent.

SAMPLE:
Prepare one liter of 1.00 M NaCl
Given:
NaCl molar mass = 58.44 g/mole

Step 1.
Weigh out 58.44 g of NaCl and transfer to a 1 L volumetric flask.

Step 2.
Dissolve the salt; fill with water to the 1-L line on the flask.

NOTE:
If you need a different M, change the mass of NaCl.

Preparing a Dilute Solution from a Stock Solution

Key: Conserve mass and moles. The molarity and volume of the stock and diluted solutions are governed by:

$$M_{stock} \times V_{stock} = M_{dilute} \times V_{dilute}$$

SAMPLE:
Prepare 50 mL of a 1.0 M solution from "a" mL of 2.0 M stock.

Given:
M_{stock} = 2.0 M
V_{stock} = a

Mdilute = 1.0 M
Vdilute = 50 mL

Calculate:
'a' mL = 50mL x 1.0M / 2.0M = 25mL

How to Calculate Colligative Properties

One example is freezing point depression:

$$\Delta T_f = - \text{ m} \times k_f \times \text{ion factor}$$

m: Molality

k_f: solvent constant

Ion factor: number of ions produced by solute: 1 for Molecular solute; 2 or more for ionic salt

---

SAMPLE:

Calculate the freezing point depression for a solution of 100g of NaCl in 500g of water.

Given:

K_f (water) = 1.86 °C/m

molar mass of salt = 58.44 g/mol

Calculate

Ion factor = 2

Mass of solvent = 500g x 1kg/1000g = 0.500 kg

Calculate

m of NaCl = $\dfrac{\left(\dfrac{100.0\text{g NaCl}}{58.44\text{g / mol}}\right)}{0.500 \text{ Kg solvent}}$ = 3.42 m

Calculate

ΔT = - 3.42 m x 1.86 °C/m x 2 = -12.72 °C

Thermodynamics: Heat, Disorder & Equilibrium

ΔH (enthalpy), ΔG (free energy) and **ΔS (entropy)** characterize a process.

$$\Delta G = \Delta H - T \, \Delta S$$
$$\Delta G = -RT \ln K_{eq}$$

How to Calculate ΔG, ΔH, ΔS from Standard Data

◆ **Given** ΔG_f° (Free Energy of Formation), in kJ/mole; ΔH_f° (Enthalpy of formation), in kJ/mole; S^0 (Standard Entropy), J/(mole K)

◆ **Calculate** ΔG = sum of product ΔG_f° - sum of reactant ΔG_f°

◆ **Calculate** ΔH = sum of product ΔH_f° - sum of reactant ΔH_f°

◆ **Calculate** ΔS = sum of product S^0 - sum of reactant S^0

SAMPLE:

ΔH for the reaction:

$CH_4(g) + 2O_2 (g) \rightarrow CO_2(g) + 2H_2O(l)$

ΔH_f°........-74.6 2 x 0.0 - 393.5 2 x -285.8

ΔH = product ΔH_f° - reactant ΔH_f°

ΔH = -393.5 -571.6 +74.6 = -890.5 kJ/mole

NOTE:
Combustion is an exothermic reaction.

SAMPLE:

ΔS for the phase change:

$\qquad H_2O(l) \rightarrow \quad H_2O(g)$

$S^0 \qquad 70.0 \qquad\quad 188.8$

ΔS = 188.8 - 70.0 = 118.8 J/(mole K)

NOTE:
A gas has more entropy than a liquid.

◆ **Does the reaction release or absorb heat? Examine ΔH.**
  • Exothermic (releases heat): $\Delta H < 0$
  • Endothermic (absorbs heat) : $\Delta H > 0$

◆ **Does the reaction proceed to completion? Is the reaction spontaneous?**

◆ **Examine ΔG**
  • $\Delta G > 0$ not spontaneous $K_{eq} < 1$
  • $\Delta G < 0$ spontaneous $K_{eq} > 1$
  • Use ΔG to calculate K_{eq}

▨ **How to Calculate K_{eq}**
$$K_{eq} = e^{-\Delta G/RT}$$

SAMPLE:
Determine K_{eq} if the ΔG of a reaction is -10 kJ/mole at 25°C.

Given: $\Delta G = -10$ kJ; T = 25°C; R = 8.3145 J/(K mol)
T(K) = 25 °C+273.15 = 298.15 K $-\Delta G$
= +10 kJ x 1,000 J/kJ = 10,000 J

$$K_{eq} = e\ \frac{10{,}000\ \text{J/mol}^{-1}}{8.314\ \text{JK}^{-1} \times 298.15\text{K}}$$

$= e^{4.03} = 56.5$
The equilibrium shifts right, a spontaneous reaction.

⚠ **Pitfall:** T must be in K; make sure you are consistent with J and kJ.

How to Handle the "Addition" of Reactions

◆ **Hess' Law:** If you "sum" reactions, you also sum ΔH, ΔG and ΔS.

SAMPLE:
Calculate ΔH for the reaction $A + D \rightarrow F$
Given: $A + B \rightarrow C$ $\Delta H = 50$ kJ/mole
Given: $C + D \rightarrow B + F$ $\Delta H = 43$ kJ/mole
Sum of the reactions gives $A + D \rightarrow F$
Calculate:
$\Delta H = 50$kJ/mole$+43$kJ/mole$=93$kJ/mole

◆ What happens if you reverse a reaction? If you reverse the reaction, change the sign of ΔH, ΔG or ΔS.

SAMPLE:
Determine ΔS for the reaction $C \rightarrow A + B$.
Given: $A + B \rightarrow C$ $\Delta S = 50$ J/mole K
For $C \rightarrow A + B$, the reverse of this reaction
Calculate $\Delta S = -50$ J/mole K

◆ How do equation coefficients impact ΔH, ΔG or ΔS? Thermodynamic properties scale with the coefficients.

SAMPLE:
Determine ΔG for the reaction:
$2A + 2C \rightarrow 2D$
Given:
$A + C \rightarrow D$ $\Delta G = -50$ kJ
For $2A + 2C \rightarrow 2D$, "double" this reaction.

Calculate: $\Delta G = 2 \times -50$ kJ $= -100$ kJ

Acid-Base Chemistry

■ **For simplicity: $[H_3O^+] = [H^+]$; [] refers to M, molarity; HAc= acetic acid; Ac^- = acetate ion**

■ **Why Water Dissociates**

$H_2O \leftrightarrow OH^- + H^+$

◆ $K_w = [OH^-][H^+] = 1 \times 10^{-14}$ at 25 ºC
◆ For pure water: $[OH^-] = [H^+] = 1 \times 10^{-7}$ M
◆ Acidic solution: $[H^+] > 1 \times 10^{-7}$ M
◆ Basic solution: $[H^+] < 1 \times 10^{-7}$ M

■ **How to Calculate pH & $[H^+]$**

$pH = - \log_{10} [H^+]$ $[H^+] = 10^{-pH}$

SAMPLE:

Determine the pH for a specific $[H^+]$
Given: $[H^+] = 1.4 \times 10^{-5}$ M
Calculate: $pH = - \log_{10} [1.4 \times 10^{-5}] = 4.85$

SAMPLE:

Determine $[H^+]$ from pH.
Given: $pH = 8.5$
Calculate: $[H^+] = 10^{-8.5} = 3.2 \times 10^{-9}$ M

How to Calculate pOH & [OH⁻]

$pOH = -\log_{10} [OH^-]$ $[OH^-] = 10^{-pOH}$

$pOH + pH = 14$ for any given solution

SAMPLE:

Determine the [OH⁻] from pOH or pH.

Given: pH = 4.5

Calculate: pOH = 14 - 4.5 = 9.5

Calculate: $[OH^-] = 10^{-9.5} = 3.2 \times 10^{-10}$ M

How to Work with K_a & K_b

◆ Acid

HA $\leftrightarrow$ H⁺ + A⁻ $K_a = \dfrac{[H^+][A^-]}{[HA]}$

• $K_a = [H^+]_{eq} [A^-]_{eq} / [HA]_{eq}$

• A⁻ : Conjugate base; $K_b(A^-) = K_w/K_a(HA)$

• $pK_a = -\log_{10} (K_a)$

• weak acids have large pK_a

◆ Base

B + H_2O $\leftrightarrow$ BH⁺ + OH⁻ $K_b = \dfrac{[OH^-][BH^+]}{[B]}$

• $K_b = [OH^-]_{eq} [BH^+]_{eq}/[B]_{eq}$

• BH⁺: Conjugate acid; $K_a(BH^+) = K_w/K_b(B)$

• $pK_b = -\log_{10} (K_b)$;

• weak bases have large pK_b

■ Calculating K_a

Substitute the experimental equilibrium concentrations into the K_a expression.

SAMPLE:
Determine K_a and pK_a from equilibrium concentration data for HA.
Given:
$[H^+]_{eq} = 1 \times 10^{-4}$ M
$[A^-]_{eq} = 1 \times 10^{-4}$ M
$[HA]_{eq} = 1.0$ M

Calculate: $K_a = \dfrac{1 \times 10^{-4} \times 1 \times 10^{-4}}{1.0} = 1 \times 10^{-8}$

Calculate: $pK_a = -\log_{10}(1 \times 10^{-8}) = 8.0$

■ How to Calculate K_b & pK_b of Conjugate Base (A⁻) of an Acid

$$K_b(A^-) = K_w/K_a(HA)$$

SAMPLE:
Determine K_b and pK_b for the acetate ion, Ac⁻.
Identify the acid: Acetic acid
$K_a(HAc) = 1.7 \times 10^{-5}$
$K_b(Ac^-) = K_w/K_a(HAc) =$

$K_b = \dfrac{1 \times 10^{-14}}{1.7 \times 10^{-5}} = 5.9 \times 10^{-10}$

$pK_b(Ac^-) = -\log_{10}(5.9 \times 10^{-10}) = 9.23$

How to Calculate Percent Dissociation of an Acid

$$\% \text{ dissociation} = 100\% \text{ x } \frac{[H^+]_{eq}}{[HA]_{initial}}$$

SAMPLE:

Determine the % dissociation for a 0.50 M HA that produces $[H^+]_{eq} = 0.10$ M and $[HA]_{eq} = 0.4$

Given: $[H^+]_{eq} = 0.10$ M; $[HA]_{initial} = 0.50$ M

Calculate: % diss. = 100% x 0.10/0.50 = 20%

⚠ **Pitfall:** Be sure to use $[HA]_{initial}$, not $[HA]_{eq}$

How to Calculate $[H^+]_{eq}$

	HA	$\leftrightarrow$	H^+	+	A^-
Init	$[HA]_{init}$		0		0
Equil	$[HA]_{init}$-a		a		a

◆ **Method:** Substitute the "Equil" expressions into K_a and solve the quadratic equation:

◆ **Start with:** $K_a = a^2 / ([HA]_{init} - a)$

◆ **This rearranges to:** $K_a \times ([HA]_{init} - a) = a^2$
$a^2 + K_a \times a - K_a \times [HA]_{init} = 0$

◆ **Given** K_a and $[HA]init$, use the quadratic formula to obtain "a", $[H^+]_{eq}$

SAMPLE:
Calculate $[H^+]$ for 0.5 M HAc
Given: $K_a = 1.7 \times 10^{-5}$
• Substitute into quadratic equation:
$a^2 + K_a \times a - K_a \times [HAc]_{init} = 0$
$a^2 + 1.7 \times 10^{-5} \times a - 1.7 \times 10^{-5} \times 0.50 = 0$
$a^2 + 1.7 \times 10^{-5} \times a - 8.5 \times 10^{-6} = 0$
• Solve the quadratic equation:
$a = [H^+]_{eq} = 0.0029$ M
• Check your work:
$K_a = (0.0029 \times 0.0029)/(0.5 - 0.0029) = 1.7 \times 10^{-5}$

⚠ **Pitfall:** Watch out for round off error when you solve for the roots of the quadratic equation.

How to Calculate $[OH^-]_{eq}$

$$B + H_2O \leftrightarrow BH^+ + OH^-$$

Init	$[B]_{init}$	0	0
Equil	$[B]_{init}-a$	a	a

◆ Method: Substitute the "Equil" expressions into K_b and solve the quadratic equation:

◆ Start with: $K_b = a^2/([B]_{init} - a)$

◆ Solve this quadratic equation for 'a' = $[OH^-]_{eq}$

■ Why Do Salts Hydrolyze?

◆ Basic Salts react with water to form OH⁻.

SAMPLE:
sodium acetate: $Ac^- + H_2O \leftrightarrow HAc + OH^-$

◆ Acidic Salts react with water to form H_3O^+

SAMPLE:
Ammonium chloride: $NH_4^+ + H_2O \leftrightarrow NH_3 + H_3O^+$

◆ Neutral Salts do not react with water.

SAMPLE:
NaCl (product of strong acid + strong base)
Na^+ and + Cl^- do not react with H_2O

How to Calculate the [H⁺] or [OH⁻] for a Salt

Step 1: Is the salt acidic, basic or neutral?

Step 2: If acidic: Identify the weak acid, and the K_a
If basic: Identify the weak base, and the K_b
If neutral: the solution will be not have acidic or basic character.

Step 3: Set up the problem as a weak base or weak acid dissociation problem.

Given the initial salt concentration, calculate the equilibrium [H⁺] or [OH⁻]

SAMPLE:
Determine the [H⁺] or [OH⁻] for a 0.40 M NaAc solution.
Step 1: NaAc is a basic salt

Step 2: Ac⁻ is the base;
$K_b(Ac^-) = K_w/K_a(HAc) = 5.9 \times 10^{-10}$

Step 3: Solve as "weak base dissociation"
$[B]_{init} = 0.40$ M Ac⁻

Calculate $[OH^-]_{equil}$

⚠ **Pitfall:** You must correctly identify the acid or base formed by the salt ions, and determine the K_a or K_b.

■ How to Calculate pH of a Buffer

Buffer of Weak Acid and Conjugate Base
Start with both weak acid, $[HA]_{init}$, and salt, $[A^-]_{init}$.
The equilibrium concentrations are governed by

$$K_a = [H^+]_{eq}[A^-]_{eq}/[HA]_{eq}$$

	HA $\leftrightarrow$	H$^+$ +	A$^-$
Init	$[HA]_{init}$	0	$[A^-]_{init}$
Equil	$[HA]_{init}$ -a	a	$[A^-]_{init}$ +a

◆ Method: Substitute the "Equil" expressions into K_a and solve the quadratic equation:

$$K_a = a \times ([A^-]_{init} + a) / ([HA]_{init} - a)$$
$$K_a \times [HA]_{init} - a \times K_a = a \times [A^-]_{init} + a^2$$
$$a^2 + a \times (K_a + [A^-]_{init}) - K_a \times [HA]_{init} = 0$$

◆ Given: K_a, $[A^-]_{init}$ and $[HA]_{init}$, solve for the roots of the quadratic, a = $[H^+]_{eq}$

SAMPLE:
Determine the pH of a buffer of 0.5 M HAc and 0.3 M Ac$^-$

Given: K_a (HAc) = 1.7 x 10^{-5}

Solve the Quadratic:
$0 = a^2 + a \times (1.7 \times 10^{-5} + 0.3) - 1.7 \times 10^{-5} \times 0.5$
$0 = a^2 + a \times (0.3) - 1.7 \times 10^{-5} \times 0.5$
$a = [H^+] = 2.8 \times 10^{-5}$ M

Calculate: pH = $- \log_{10} (2.8 \times 10^{-5}) = 4.55$

How to Use the Henderson-Hassellbach Approximation for Buffer pH

◆ Assume that "a" in the previous problem
is $\ll [HAc]_{init}$ and $[Ac^-]_{init}$

Henderson Hassellbach: $pH = pK_a + \log_{10} \dfrac{[A^-]}{[HA]}$

SAMPLE:

Examine previous buffer problem:

Given: $[Ac^-]$

$[A^-] = [Ac^-] = 0.3$ M; $[HA] = [HAc] = 0.5$ M

$pK_a = pK_a(HAc) = 4.77$

Calculate:

$pH = 4.77 + \log_{10}(0.3/0.5) = 4.77 - 0.22 = 4.55$

The Henderson Hassellbach approximation works.

How to Do an Acid-Base Titration

A systematic acid-base neutralization used to determine the concentration of an unknown acid or base. At the equivalence point

moles of acid = moles of base.

SAMPLE:

The titration of 50.00 mL of an HCl solution requires 25.00 mL of 1.00 M NaOH.
Calculate the [HCl].

Equation: $HCl + NaOH \rightarrow NaCl + H_2O$
This gives a 1:1 molar ratio of HCl: NaOH

At the equivalence point: the moles balance, or more conveniently:
mmoles HCl = mmoles NaOH.
M(HCl) x vol-acid (mL)
= vol-base (mL) x M (NaOH)
M(HCl)
= vol-base (mL) x M (NaOH) / vol-acid (mL)

M(HCl) = 25.00 mL x 1.00 M /50.00 mL
= 0.50 M HCl

⚠ **Pitfall:** Watch the units on volume and molarity, work with "L & mole," or "mL & mmole."

Examination of Chemical Equilibrium

▮ Incomplete Reactions

◆ For a reaction that has not gone to completion:

$$a\,A + b\,B \leftrightarrow c\,C + d\,D$$

◆ At equilibrium, the process is described by the **equilibrium constant**, K_c.

$$K_c = \frac{[C]_{eq}^{c}\,[D]_{eq}^{d}}{[A]_{eq}^{a}\,[B]_{eq}^{b}}$$

◆ For all other conditions, the process is described by the **reaction quotient**, Q_c:

$$Q_c = \frac{[C]^{c}\,[D]^{d}}{[A]^{a}\,[B]^{b}}$$

if $Q_c = K_c$, the reaction is at equilibrium

if $Q_c > K_c$, the reaction will shift to the left

if $Q_c < K_c$, the reaction will shift to the right

Gas phase reactions may be described with K_p, based on reagent partial pressures. These calculations follow the same strategy as K_c.

■ How to Determine If the Reaction Is at Equilibrium
  ◆ Compare Q_c with K_c

SAMPLE:
For the reaction: A ↔ C, K_c = 0.60; the observed
[A] = 0.1 and [C] = 0.20.
Is the reaction at equilibrium?

If not, predict the shift.

Calculate: Q_c = [C]/[A] = 0.20/0.10 = 2.0
Given: K_c = 0.60
$Q_c > K_c$; process is not at equilibrium the reaction
        will shift to the left.

■ **How to Predict Equilibrium Concentrations**

SAMPLE:
Calculate the equilibrium concentrations for the
following gas phase reaction data:
$CO + H_2O \rightarrow CO_2 + H_2$

Given:
K_c = 0.64; $[CO]_{init} = [H_2O]_{init}$ = 0.5 M:

	$CO(g) +$	$H_2O(g) \leftrightarrow$	$CO_2(g) +$	$H_2(g)$
Init	$[CO]_{init}$	$[H_2O]_{init}$	0	0
Equil	$[CO]_{init}$ -a	$[H_2O]_{init}$ -a	a	a

NOTE:
The change "a" is the same for each because of the 1:1:1:1 coefficients in the equation.

SAMPLE:
Identify the equilibrium expression:
$$K_c = \frac{[CO_2]_{eq}{}^c\,[H_2]_{eq}{}^d}{[CO]_{eq}{}^a\,[H_2O]_{eq}{}^b}$$

Substitute "equil" values:
$K_c = a^2/\{([CO]_{init} -a) \times ([H_2O]_{init} -a)\}$
$0.64 = a^2/\{(0.50 -a)(0.50 -a)\}$

Take square root of each side:
$0.8 = a/(0.5-a)$ or $-0.8 = a / (0.5-a)$
$a = 0.222$ or $a = -1.2$

Use the first option, since "a" must be positive.
$[CO_2]_{eq} = [H_2]_{eq} = 0.222$ M
$[CO]_{eq} = [H_2O]_{eq} = 0.50 - 0.222 = 0.278$ M

Check your work:
$K_c = (0.222^2)/ (0.278^2) = 0.64$

⚠ **Pitfall:** Watch out for round-off error. Take the root that gives positive concentrations.

How to Determine the Solubility Limit from K_{sp} (in moles/L or g/L)

SAMPLE:
Determine the solubility limit for silver chloride, AgCl, given $K_{sp} = 1.77 \times 10^{-10}$

Given: $AgCl \text{ (s)} \leftrightarrow Ag^+ \text{ (aq)} + Cl^- \text{ (aq)}$;
$K_{sp} = [Ag^+][Cl^-] = 1.77 \times 10^{-10}$
Given: AgCl molar mass = 143.32 g/mole
At equilibrium, $[Ag^+]_{eq} = [Cl\text{-}]_{eq} = \sqrt{K_{sp}}$

Calculate: $[Ag^+]_{eq} = \sqrt{(1.77 \times 10^{-10})}$
$= 1.33 \times 10^{-5}$ M AgCl
This is also $[AgCl]_{eq}$, the molar solubility limit for AgCl.

Calculate: The AgCl g/L solubility limit
$= [AgCl]_{eq}$ x molar mass of AgCl
$= 1.33 \times 10^{-5}$ moles/L AgCl x 143.32 g/mole
$= 1.9 \times 10^{-3}$ g/L

Kinetics & Mechanisms

The goal of a kinetic study is to measure the reaction rate, determine the rate constant, k, the rate law and the activation energy (E_a).

How to Determine the Reaction Rate

For "A→B," the reaction rate is the rate of appearance of product, $\Delta[B]/\Delta time$; or the rate of loss of reactant: $-\Delta[A]/\Delta time$.

SAMPLE:
How would you characterize the rate of:
$CaCO_3$ (s) → CaO (s)+ CO_2 (g)
Answer: Focus on rate of CO_2 production;
Rate = $\Delta[CO_2]/\Delta time$.

How to Determine the Rate Law

The rate law gives the order of the reaction based on the steps in the overall reaction "A + B → C."

◆ Rate = k [A], for a first-order reaction.
◆ Rate = k [A]², or k [A][B] for a second order reaction.
◆ Rate = k [A]⁰, for a zero-order reaction.

⚠ **Pitfall:** Equation coefficients describe the balanced overall reaction, not the mechanism and rate-law.

■ **Two Common Rate-Law Methods:** "initial rates" and "integrated rate equations."

Consider the reaction "A→B", with a rate law of the form: rate = k $[A]^x$

The goal of a kinetic study is to determine "x", the order of the reaction.

■ **Initial Rate Strategy:**
◆ **Step 1:** For $[A]_1$, measure the time required to produce $\Delta[B]$ of product.
◆ **Calculate:** $Rate_1 = \Delta[B]/time$
◆ **Step 2:** Measure the new reaction rate, $rate_2$, for a different concentration, $[A]_2$

The interplay of rate, [A] and x are governed by the equation: $\dfrac{Rate\ 1}{Rate\ 2} = \left(\dfrac{[A]_1}{[A]_2}\right)^x$

SAMPLE:
Determine "x," if doubling [A] also doubles the rate:
$Rate_1/Rate_2 = 2$
$[A]_1/[A]_2 = 2$

$2 = 2^x$, x=1
this is a 1st order process

SAMPLE:
Determine "x" if doubling [A] increases the rate by 4-fold:
$Rate_1/Rate_2 = 4$
$[A]_1/[A]_2 = 2$

$4 = 2^x$, $x = 2$
this is a 2nd order process

Integrated Rate Equation Strategy
Analyze "[A] vs. time" data for the reaction.

◆ The reaction is 1st order in [A] if the graph "ln [A] vs. t" is linear.

◆ The reaction is 2nd order in [A] if the graph "1/[A] vs t" is linear.
In each case, k is the slope of the line.

◆ If the reaction is second order, k [A][B], the data exhibits a specific mathematical form.

▦ How to Determine the Activation energy, E_a

$$\ln \frac{k_2}{k_1} = \frac{-E_a}{R}\left(\frac{1}{T_1} - \frac{1}{T_2}\right)$$

▦ Applications

◆ Predict k_1 at T_1, given E_a and k_2 at T_2

◆ Determine E_a from k_1, k_2, T_1 and T_2

SAMPLE:

The rate constant doubles when the temperature changes from 25.0°C to 50.0°C.

What is E_a?

Given: $k_1/k_2 = 2$

Calculate: $T_1 = 50.0°C + 273.15 = 323.2$ K

Calculate: $T_2 = 25.0°C + 273.15 = 298.2$ K

Calculate: $\Delta T = T_1 - T_2 = 25.0$ K

Calculate: $E_a = R \ln\left(\frac{k_1}{k_2}\right) \frac{T_1 T_2}{\Delta T}$

$= 8.314$ J/mole K x ln (2) x 323.2 K
x 298.2 K / 25.0 K

$E_a = 22,200$ J/mole $= 22.2$ kJ/mole

⚠ **Pitfall:** T must be in Kelvin; if you use the equation with "$1/T_1 - 1/T_2$" beware of round off error in calculating inverse T.

Kinetics & Equilibrium

An equilibrium is characterized by competing forward and reverse reactions.

◆ The forward and reverse rate constants (k_f & k_r) are related to the equilibrium constant, K_{eq}

◆ At equilibrium: the forward and reverse reaction rates are equal.

As a result $K_{eq} = k_f / k_r$

17 Chem Lab Basics

General Lab Guidelines

■ Always work with instructor **supervision.**

■ Always wear **goggles** in the lab, even over eyeglasses; replace contact lenses with eyeglasses.

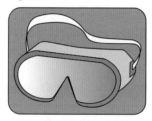

■ Wear an **apron, lab-coat** and **gloves** to limit your chemical exposure and to save clothing from chemical stains.

■ Wear **closed-toe** shoes and long pants to protect your feet and legs.

■ Tie back hair and avoid bulky sleeves which **interfere** with work.

■ Food and drink **should not** be in the lab.

■ **Wash** your hands after each session, before leaving the lab.

Safe Use of Lab Equipment

■ Lab equipment is delicate and expensive; learn to use it correctly; ask for assistance if you need help.

■ Do not use worn or frayed electrical cords.

■ Be aware of the risk of static electricity-it may harm computers and can ignite flammable solvents.

■ Watch out for chipped or cracked glassware; discard in the glass-recycle box.

■ **Thermometer:** Use "non-mercury" for routine work

■ **Refrigerator:** Store chemicals in sealed containers; do not store food with chemicals.

■ **Compressed-gas cylinders:** Secure to a wall or bench; falling cylinders cause serious injuries.

■ **Types of Material**
   ◆ Plastic (melts if heated; may dissolve in acetone)
   ◆ Borosilicate glass (Pyrex, high temp)
   ◆ Flint-glass (for room temp)
   ◆ Ceramic (high temp; stain resistant); crucible, clay triangle
   ◆ Metal (high temperature)

Working with Chemicals

■ Heating Labware

◆ Use tongs to handle labware while it is heated by a burner or hotplate.

◆ Allow the item to cool to room temperature before weighing.

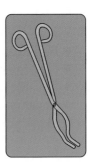

■ For Liquid Reagent

◆ Cover the beaker with a watch glass

◆ Use "boiling stones" to promote smooth boiling

◆ Flammable solvent: take care when heating with hot plate; avoid use of gas burner

◆ Handle test tube with wire-holder

■ For Solid Reagent

◆ Use a weighing dish on the balance

◆ Cover the dish to prevent loss, spills or contamination

Useful Chemical Information

■ Liquid Solubility rule: "like dissolves like"
■ Water "The Universal Solvent"
- ◆ Boiling pt: . . . . .100.00 ºC
- ◆ Freezing pt: . . . . 0.00 ºC
- ◆ Density: . . . . 1.00 g/mL at 4 ºC
- ◆ Molar mass: . . . . 18.015 g
- ◆ Vapor pressure: . . . . . . 23.8 mm Hg, 25 ºC.

■ **Organic Compounds: General Rules**
- ◆ Non-polar (alkane, oil, benzene) are soluble in non-polar organic solvent, but insoluble in water.
- ◆ Polar compounds (amine, alcohol, organic acid) tend to dissolve in water.

■ **Organic Solvents:**

	T_b	P	M	
◆ ethanol	79	0.79	46.07	Polar
◆ methanol	65	0.79	32.04	Polar
◆ acetone	56	0.79	58.08	Polar
◆ isopropanol	82	0.79	60.11	Polar
◆ benzene	80	0.88	78.12	Nonpolar
◆ toluene	111	0.87	92.15	Nonpolar

T_b = Boiling pt (ºC)

P = Density (g/mL)

M = molar mass (g/mole)

■ Acids

	Commercial	Reagent
Hydrochloric, HCl	11.6 M	Pungent
Nitric, HNO_3	16.0 M	Oxidizer
Sulfuric, H_2SO_4	18.0 M	Dehydrating Agent
Acetic	6.27 M	
Glacial acetic	17.4 M	
Phosphoric	14.7 M	

■ Bases

NaOH and KOH, hygroscopic pellets

	Commercial	Reagent
NaOH	19.1 M	
Aqueous ammonia	14.8 M	pungent

■ Dangerous Chemical Combinations

* Acid + base: react exothermically
* CN^- + acid => HCN   deadly gas
* S^{2-} + acid => H_2S poisonous gas

Data Manipulation

■ Lab Units & Conversion Factors

All data has a "number" and a "unit"

◆ Mass (gram, g) (1000 g = 1 kg)

◆ Time (second, s) (60 s = 1 minute)

◆ Length (meter, m) (1 m = 100 cm)

◆ Volume (liter, L) (1000 mL = 1 L)

◆ Temperature (°C, Centigrade)

   Fahrenheit: °F = °C (9/5) + 32

   Kelvin: K = °C + 273.15

◆ Pressure: (pascal, Pa)

   760 mm Hg = 1 atm = 14.70 lb/in²

   1 atm = 101,325 Pa = 1.01325 bar

■ Significant Figures (sig. figs.)

◆ Record the number of digits appropriate for the measuring device, plus record one "approximate" digit. Exponents are always significant.

◆ Add/subtract: For final answer: number of decimal places is given by datum with least number of decimal places.

◆ Multiply/divide: For the final answer: number of sigfigs is given by the datum with the least number of sigfigs.

■ Graphing (x,y) Data

Set range to use all of the graph page; label axes and clearly mark data points

◆ **Equation for a line (x,y):**

   y = mx + b (m = slope, b = intercept)

◆ **Average or Mean Value:** sum all data values and divide by the number of data points

Preparing a Solution

Use volumetric glassware; add reagent, dissolve in some solvent and then dilute to the "mark" on the flask with additional solvent.

■ Dilute solution from stock reagent: select volume, v-dil, and desired concentration, c-dil; use " v-stock" of reagent of concentration "c-stock". Use the equation: v-dil x c-dil = v-stock x c-stock

■ From pure reagent: select the desired concentration and volume; determine the required # of moles, then calculate the mass (using the molar mass).

■ **Dilution of Acid or Base**
   ◆ Always add acid (or base) to water, slowly, with stirring. Heat is produced in the process.

■ **Inorganic Salts**
   ◆ **Soluble:** acetate, nitrate, alkali metal (Na, K, Li, Rb, Cs), ammonium, perchlorate
   ◆ **Mostly soluble:** chloride, bromide, iodide (except Ag, Pb, Hg(I)); sulfate (except Ba, Pb and Hg(I))
   ◆ **Mostly insoluble:** carbonate, hydroxide, oxide, sulfide, phosphate, chromate; (except for "soluble")

■ **pH and Acid/Base Concentration:**
   ◆ pH = - $\log_{10}$ [H^+] ; molar concentration
   ◆ Base turns red-litmus blue;
   ◆ Acid turns blue-litmus red
   ◆ **Acidic salt:** NH_4Cl (from weak base + strong acid)
   ◆ **Basic salt:** NaCN (from strong base + weak acid)
   ◆ **Neutral salt:** NaCl (from strong acid + strong base)

NFPA Hazard Codes (National Fire Prevention Association)
(Highlights major chemical hazards)

Know Your Lab Reagents

Some chemicals are toxic; all can cause harm if used incorrectly. Learn about reagents **before** using them in an experiment. Read your lab manual and textbook, talk to your instructor; if in doubt, **ask questions!**

Material Safety Data Sheet (MSDS) gives a description of the hazard a substance may pose.

Chemical Storage Codes

Chemicals in the same color group can normally be stored together; exceptions noted on the label

◆ Health Hazard
◆ Reactive and Oxidizing
◆ Flammable
◆ Corrosive
◆ Minimal Hazard

Exposure to Chemicals

While working in the lab you will use a number of reagents, giving ample chance for exposure to the harmful effects of chemicals.

■ **Possible Risks**
- ◆ Inhaling chemical powder or vapor
- ◆ Ingesting solid or liquid chemicals by mouth
- ◆ Puncturing skin with a sharp-object and injecting chemicals into your body
- ◆ Absorbing chemicals through your skin

Chemical Spills

■ **On the Floor or Benchtop**
- ◆ For **small spill:** wear gloves, neutralize acid/base; absorb using paper-towels and discard in a labeled bag.
- ◆ **Larger spill:** Notify the instructor; wear gloves and shoe protectors, use a spill-kit designed for the chemical.
- ◆ Clean up all spills **promptly** to prevent further accidents.

■ **On Your Clothing or Skin**
- ◆ Remove affected article of clothing; wash exposed skin with water and apply first-aid. Treat promptly to minimize harm.
- ◆ If a large area is exposed, use the safety shower, then apply first aid.

Safety Training

■ Safety is an integral part of working in the chemistry laboratory, and a responsibility shared by students and instructors.

■ Learning about safety is part of your education; skills you gain in the lab will serve you in future careers and in life...if nothing else they will make you a better cook!

■ **Be Prepared** - Where is the. . .
   ◆ **Lab exit** - Know how to get out fast in an emergency
   ◆ **Nearest phone** - Dial 911 or local emergency number
   ◆ **Fume hood** - Use for any noxious reagent
   ◆ **Eyewash station and safety shower** - For washing skin or eyes exposed to chemicals
   ◆ **Fire-extinguisher** - Use to douse small fires

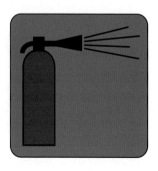

■ **Personal Responsibility:**

◆ **Rule 1:** Protect yourself! Your mistakes will likely harm you more than anyone else.

◆ **Rule 2:** Read the lab manual before class. Come to lab prepared to work on the assigned experiment.

◆ **Rule 3:** Always pay attention as you work. Watch other students; you are impacted by their mistakes.

◆ **Rule 4:** Clean up your own mess...You are a partner in maintaining a safe lab.

◆ Keep your work space clean and organized.

◆ Wash labware with detergent; rinse with deionized or distilled water; use a wash bottle to conserve water; drain excess liquid, allow object to dry before storing.

◆ Shared-equipment: wash before and after each use.

◆ After each lab session, return reagents and equipment to the designated storage areas.

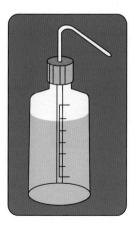

First Aid

■ Check with instructor for local guidelines.

◆ **Burn from hot labware:**
Minor: apply cold water
Serious: contact medical help

◆ *Cut from broken glassware:*
Minor: Wash with soap, apply antiseptic and sterile bandage.
Serious: Control bleeding by applying pressure with sterile pad, contact emergency medical help.

◆ **Skin-exposure to a chemical:**
Rinse with water; if condition develops, contact medical personnel.

◆ **Feel lightheaded (or passing out):**
Move affected person to fresh air outside of the lab; contact medical personnel if the condition persists.

◆ **Burning clothing:**
Do not panic, drop to the floor and smother the flame; use safety shower to treat burn; contact emergency medical personnel.

Waste Management

◆ Follow the instructor's directions for disposal of all lab materials. Most chemicals should not be poured down the drain.

◆ All toxic metals and halogenated solvents must be collected for proper disposal.

Waste Prevention: Use only the required amount of reagent; excess material cannot be returned to reagent jar; it is "waste". A spot plate is an excellent means to conserve reagents.